COURSE OF THEORETICAL PHYSICS

Volume 1

MECHANICS

THIRD EDITION

COURSE OF THEORETICAL PHYSICS

Vol. 2. *The Classical Theory of Fields*

Vol. 3. *Quantum Mechanics—Non-relativistic Theory*

Vol. 4. *Relativistic Quantum Theory*

Vol. 5. *Statistical Physics*

Vol. 6. *Fluid Mechanics*

Vol. 7. *Theory of Elasticity*

Vol. 8. *Electrodynamics of Continuous Media*

Vol. 9. *Physical Kinetics*

A SHORTER COURSE OF THEORETICAL PHYSICS
(Based on the Course of Theoretical Physics)

Vol. 1. *Mechanics and Electrodynamics*

Vol. 2. *Quantum Mechanics*

L. D. LANDAU

MECHANICS

THIRD EDITION

by

L. D. LANDAU AND E. M. LIFSHITZ

INSTITUTE OF PHYSICAL PROBLEMS, U.S.S.R. ACADEMY OF SCIENCES

Volume 1 of *Course of Theoretical Physics*

Translated from the Russian by
J. B. SYKES AND J. S. BELL

PERGAMON PRESS

OXFORD · NEW YORK · TORONTO
SYDNEY · PARIS · FRANKFURT

U.K.	Pergamon Press Ltd., Headington Hill Hall, Oxford OX3 0BW, England
U.S.A.	Pergamon Press Inc., Maxwell House, Fairview Park, Elmsford, New York 10523, U.S.A.
CANADA	Pergamon of Canada Ltd., P.O. Box 9600, Don Mills, M3C 2T9, Ontario, Canada
AUSTRALIA	Pergamon Press (Aust.) Pty. Ltd., 19a Boundary Street, Rushcutters Bay, N.S.W. 2011, Australia
FRANCE	Pergamon Press SARL, 24 rue des Ecoles, 75240 Paris, Cedex 05, France
WEST GERMANY	Pergamon Press GmbH, 6242, Kronberg-Taunus, Pferdstrasse 1, Frankfurt-am-Main, West Germany

Copyright © 1960, 1969 and 1976 Pergamon Press Ltd.

First English edition 1960

Second edition 1969

Third edition 1976

Library of Congress Catalog Card No. 79–87907

Translated from the 3rd revised and enlarged edition of Mekhanika by L. D. Landau and E. M. Lifshitz. Nauka, Moscow, 1973

Library of Congress Cataloging in Publication Data

Landau, Lev Davidovich, 1908–1968.
Mechanics.

(Courses of theoretical physics; v. 1)
Translation of Mekhanika.
"Lev Davidovich Landau (1908–1968)" by E. M. Lifshitz
Lifshitz: p.
Includes bibliographical references and index.
1. Mechanics, Analytic. I. Lifshits, Evgenii
Mikhailovich, joint author. II. Title.
QA805.L283 1976   531'.01'515   76–18997
ISBN 0–08–021022–8
ISBN 0–08–021020–1 flexi

Printed in Great Britain by Biddles Ltd., Guildford, Surrey

ISBN 0 08 021022 8 Hard

CONTENTS

V. SMALL OSCILLATIONS

VI. MOTION OF A RIGID BODY

VII. THE CANONICAL EQUATIONS

PREFACE TO THE THIRD ENGLISH EDITION

THIS book continues the series of English translations of the revised and augmented volumes in the *Course of Theoretical Physics*, which have been appearing in Russian since 1973. The English translations of volumes 2 (*Classical Theory of Fields*) and 3 (*Quantum Mechanics*) will shortly both have been published. Unlike those two, the present volume 1 has not required any considerable revision, as is to be expected in such a well-established branch of theoretical physics as mechanics is. Only the final sections, on adiabatic invariants, have been revised by L. P. Pitaevskiĭ and myself.

The *Course of Theoretical Physics* was initiated by Landau, my teacher and friend. Our work together on these books began in the late 1930s and continued until the tragic accident that befell him in 1962. Landau's work in science was always such as to display his striving for clarity, his effort to make simple what was complex and so to reveal the laws of nature in their true simplicity and beauty. It was this aim which he sought to instil into his pupils, and which has determined the character of the *Course*. I have tried to maintain this spirit, so far as I was able, in the revisions that have had to be made without Landau's participation. It has been my good fortune to find a colleague for this work in L. P. Pitaevskiĭ, a younger pupil of Landau's.

The present edition contains the biography of Landau which I wrote in 1969 for the posthumous Russian edition of his *Collected Works*. I should like to hope that it will give the reader some slight idea of the personality of that remarkable man.

The English translations of the *Course* were begun by Professor M. Hamermesh in 1951 and continued by Dr. J. B. Sykes and his colleagues. No praise can be too great for their attentive and careful work, which has contributed so much to the success of our books in the English-speaking world.

Institute of Physical Problems　　　　　　　　　　　　　E. M. LIFSHITZ
U.S.S.R. Academy of Sciences
Moscow

LEV DAVIDOVICH LANDAU (1908–1968)†

VERY little time has passed since the death of Lev Davidovich Landau on 1 April 1968, but fate wills that even now we view him at a distance, as it were. From that distance we perceive more clearly not only his greatness as a scientist, the significance of whose work becomes increasingly obvious with time, but also that he was a great-hearted human being. He was uncommonly just and benevolent. There is no doubt that therein lie the roots of his popularity as a scientist and teacher, the roots of that genuine love and esteem which his direct and indirect pupils felt for him and which were manifested with such exceptional strength during the days of the struggle to save his life following the terrible accident.

To him fell the tragic fate of dying twice. The first time it happened was six years earlier on 7 January 1962 when on the icy road, en route from Moscow to Dubna, his car skidded and collided with a lorry coming from the opposite direction. The epic story of the subsequent struggle to save his life is primarily a story of the selfless labour and skill of numerous physicians and nurses. But it is also a story of a remarkable feat of solidarity. The calamitous accident agitated the entire community of physicists, arousing a spontaneous and instant response. The hospital in which Landau lay unconscious became a centre to all those – his students and colleagues – who strove to make whatever contributions they could to help the physicians in their desperate struggle to save Landau's life.

"Their feat of comradeship commenced on the very first day. Illustrious scientists who, however, had no idea of medicine, academicians, corresponding members of the scientific academies, doctors, candidates, men of the same generation as the 54-year-old Landau as well as his pupils and *their* still more youthful pupils – all volunteered to act as messengers, chauffeurs, intermediaries, suppliers, secretaries, members of the watch and, lastly, porters and labourers. Their spontaneously established headquarters was located in the office of the Physician-in-Chief of Hospital No. 50 and it became a round-the-clock organizational centre for an unconditional and immediate implementation of any instruction of the attending physicians.

---

† By E. M. Lifshitz; written for the Russian edition of Landau's *Collected Papers*, and first published in Russian in *Uspekhi fizicheskikh nauk* **97,** 169-183, 1969. This translation is by E. Bergman (first published in *Soviet Physics Uspekhi* **12,** 135-143, 1969), with minor modifications, and is reprinted by kind permission of the American Institute of Physics. The reference numbers correspond to the numbering in the *Collected Papers of L. D. Landau* (Pergamon Press, Oxford 1965).

"Eighty-seven theoreticians and experimenters took part in this voluntary rescue team. An alphabetical list of the telephone numbers and addresses of any one and any institution with which contact might be needed at any instant was compiled, and it contained 223 telephone numbers! It included other hospitals, motor transport bases, airports, customs offices, pharmacies, ministries, and the places at which consulting physicians could most likely be reached.

"During the most tragic days when it seemed that 'Dau is dying' – and there were at least four such days – 8–10 cars could be found waiting at any time in front of the seven-storey hospital building. . . .

"When everything depended on the artificial respiration machine, on 12 January, a theoretician suggested that it should be immediately constructed in the workshops of the Institute of Physical Problems. This was unnecessary and naive, but how amazingly spontaneous! The physicists obtained the machine from the Institute for the Study of Poliomyelitis and carried it in their own hands to the ward where Landau was gasping for breath. They saved their colleague, teacher, and friend.

"The story could be continued without limit. This was a real fraternity of physicists. . . ."†

And so, Landau's life was saved. But when after three months he regained consciousness, it was no longer the same man whom we had known. He was not able to recover from all the consequences of his accident and never again completely regained his abilities. The story of the six years that followed is only a story of prolonged suffering and pain.

*　　　*　　　*

Lev Davidovich Landau was born on 22 January 1908 in Baku, in the family of a petroleum engineer who worked on the Baku oil-fields. His mother was a physician and at one time had engaged in scientific work on physiology.

He completed his school course at the age of 13. Even then he already was attracted by the exact sciences, and his mathematical ability manifested itself very early. He studied mathematical analysis on his own and later he used to say that he hardly remembered a time when he did not know differentiation and integration.

His parents considered him too young to enter a university and for a year he attended the Baku Economic Technicum. In 1922 he enrolled at Baku University where he studied simultaneously in two departments: Physico-mathematical and Chemical. Subsequently he did not continue his chemical education but he remained interested in chemistry throughout his life.

In 1924 Landau transferred to the Physics Department of Leningrad

---

† From D. Danin, "Comradeship", *Literaturnaya Gazeta* (*Literary Gazette*), 21 July 1962.

University. In Leningrad, the main centre of Soviet physics at that time, he first made the acquaintance of genuine theoretical physics, which was then going through a turbulent period. He devoted himself to its study with all his youthful zeal and enthusiasm and worked so strenuously that often he became so exhausted that at night he could not sleep, still turning over formulae in his mind.

Later he used to describe how at that time he was amazed by the incredible beauty of the general theory of relativity (sometimes he even would declare that such a rapture on first making one's acquaintance with this theory should be a characteristic of any born theoretical physicist). He also described the state of ecstasy to which he was brought on reading the articles by Heisenberg and Schrödinger signalling the birth of the new quantum mechanics. He said that he derived from them not only delight in the true glamour of science but also an acute realization of the power of the human genius, whose greatest triumph is that man is capable of apprehending things beyond the pale of his imagination. And of course, the curvature of space-time and the uncertainty principle are precisely of this kind.

In 1927 Landau graduated from the university and enrolled for postgraduate study at the Leningrad Physicotechnical Institute where even earlier, in 1926, he had been a part-time research student. These years brought his first scientific publications. In 1926 he published a theory of intensities in the spectra of diatomic molecules [1],† and as early as 1927, a study of the problem of damping in quantum mechanics, which first introduced a description of the state of a system with the aid of the density matrix.

His fascination with physics and his first achievements as a scientist were, however, at the time beclouded by a painful diffidence in his relations with others. This trait caused him a great deal of suffering and at times – as he himself confessed in later years – led him to despair. The changes which occurred in him with the years and transformed him into a buoyant and gregarious individual were largely a result of his characteristic self-discipline and feeling of duty toward himself. These qualities, together with his sober and self-critical mind, enabled him to train himself and to evolve into a person with a rare ability – the ability to be happy. The same sobriety of mind enabled him always to distinguish between what is of real value in life and what is unimportant triviality, and thus also to retain his mental equilibrium during the difficult moments which occurred in his life too.

In 1929, on an assignment from the People's Commissariat of Education, Landau travelled abroad and for one and a half years worked in Denmark, Great Britain and Switzerland. To him the most important part of his trip was his stay in Copenhagen where, at the Institute of Theoretical Physics,

---

† He did not know, however, at the time that these results had been already published a year earlier by Hönl and London.

theoretical physicists from all Europe gathered round the great Niels Bohr and, during the famous seminars headed by Bohr, discussed all the basic problems of the theoretical physics of the time. This scientific atmosphere, enhanced by the charm of the personality of Bohr himself, decisively influenced Landau in forming his own outlook on physics and subsequently he always considered himself a disciple of Niels Bohr. He visited Copenhagen two more times, in 1933 and 1934. Landau's sojourn abroad was the occasion, in particular, of his work on the theory of the diamagnetism of an electron gas [4] and the study of the limitations imposed on the measurability of physical quantities in the relativistic quantum region (in collaboration with Peierls) [6].

On his return to Leningrad in 1931 Landau worked in the Leningrad Physicotechnical Institute and in 1932 he moved to Khar'kov, where he became head of the Theoretical Division of the newly organized Ukrainian Physicotechnical Institute, an offshoot of the Leningrad Institute. At the same time he headed the Department of Theoretical Physics at the Physics and Mechanics Faculty of the Khar'kov Mechanics and Machine Building Institute and in 1935 he became Professor of General Physics at Khar'kov University.

The Khar'kov period was for Landau a time of intense and varied research activity.† It was there that he began his teaching career and established his own school of theoretical physics.

Twentieth-century theoretical physics is rich in illustrious names of trail-blazing creators, and Landau was one of these creators. But his influence on scientific progress was far from exhausted by his personal contribution to it. He was not only an outstanding physicist but also a genuinely outstanding educator, a born educator. In this respect one may take the liberty of comparing Landau only to his own teacher – Niels Bohr.

The problems of the teaching of theoretical physics as well as of physics as a whole had first attracted his interest while still quite a young man. It was there, in Khar'kov, that he first began to work out programmes for the "theoretical minimum" – programmes of the basic knowledge in theoretical physics needed by experimental physicists and by those who wish to devote themselves to professional research work in theoretical physics. In addition to drafting these programmes, he gave lectures on theoretical physics to the scientific staff at the Ukrainian Physicotechnical Institute as well as to students of the Physics and Mechanics Faculty. Attracted by the ideas of reorganizing instruction in physics as a whole, he accepted the Chair of General Physics at Khar'kov State University (and subsequently, after

---

† The extent of Landau's scientific activities at the time can be grasped from the list of studies he completed during the year 1936 alone: theory of second-order phase transitions [29], theory of the intermediate state of superconductors [30], the transport equation in the case of Coulomb interaction [24], the theory of unimolecular reactions [23], properties of metals at very low temperatures [25], theory of the dispersion and absorption of sound [22, 28], theory of photoelectric effects in semiconductors [21].

the war, he continued to give lectures on general physics at the Physicotechnical Faculty of Moscow State University).

It was there also, in Khar'kov, that Landau had conceived the idea and
began to implement the programme for compiling a complete Course of
Theoretical Physics and Course of General Physics. All his life long, Landau
dreamed of writing books on physics at every level – from school textbooks
to a course of theoretical physics for specialists. In fact, by the time of his
fateful accident, nearly all the volumes of the *Course of Theoretical Physics*
and the first volumes of the *Course of General Physics* and *Physics for
Everyone* had been completed. He also had drafted plans for the compilation
of textbooks on mathematics for physicists, which should be "a guide to
action", should instruct in the practical applications of mathematics to
physics, and should be free of the rigours and complexities unnecessary to
this course. He did not have time to begin to translate this programme into
reality.

Landau always attached great importance to the mastering of mathematical techniques by the theoretical physicist. The degree of this mastery
should be such that, insofar as possible, mathematical complications would
not distract attention from the physical difficulties of the problem – at least
whenever standard mathematical techniques are concerned. This can be
achieved only by sufficient training. Yet experience shows that the current
style and programmes for university instruction in mathematics for physicists often do not ensure such training. Experience also shows that after a
physicist commences his independent research activity he finds the study
of mathematics too "boring".

Therefore, the first test which Landau gave to anyone who desired to
become one of his students was a quiz in mathematics in its "practical"
calculational aspects.† The successful applicant could then pass on to the
study of the seven successive sections of the programme for the "theoretical
minimum", which includes basic knowledge of all the domains of theoretical
physics, and subsequently take an appropriate examination. In Landau's
opinion, this basic knowledge should be mastered by any theoretician
regardless of his future specialization. Of course, he did not expect anyone
to be as universally well-versed in science as he himself. But he thus
manifested his belief in the integrity of theoretical physics as a single
science with unified methods.

At first Landau himself gave the examination for the "theoretical
minimum". Subsequently, after the number of applicants became too large,
this duty was shared with his closest associates. But Landau always re-

---

† The requirements were: ability to evaluate any indefinite integral that can be expressed
in terms of elementary functions and to solve any ordinary differential equation of the standard
type, knowledge of vector analysis and tensor algebra as well as of the principles of the theory
of functions of a complex variable (theory of residues, Laplace method). It was assumed that
such fields as tensor analysis and group theory would be studied together with the fields of
theoretical physics to which they apply.

served for himself the first test, the first meeting with each new young applicant. Anyone could meet him – it was sufficient to ring him up and ask him for an interview.

Of course, not every one who began to study the "theoretical minimum" had sufficient ability and persistence to complete it. Altogether, between 1934 and 1961, 43 persons passed this test. The effectiveness of this selection can be perceived from the following indicative facts alone: of these persons 7 already have become members of the Academy of Sciences and an additional 16, doctors of sciences.

In the spring of 1937 Landau moved to Moscow where he became head of the Theoretical Division of the Institute of Physical Problems which had not long before been established under the direction of P. L. Kapitza. There he remained to the end of his life; in this Institute, which became a home to him, his varied activity reached its full flowering. It was there, in a remarkable interaction with experimental research, that Landau created what may be the outstanding accomplishment of his scientific life – the theory of quantum fluids.

It was there also that he received the numerous outward manifestations of the recognition of his contributions. In 1946 he was elected a full Member of the USSR Academy of Sciences. He was awarded a number of orders (including two Orders of Lenin) and the honorific title of Hero of Socialist Labour – a reward for both his scientific accomplishments and his contribution to the implementation of important practical State tasks. He was awarded the State Prize three times and in 1962, the Lenin Prize. There also was no lack of honorific awards from other countries. As far back as 1951 he was elected member of the Danish Royal Academy of Sciences and in 1956, member of the Netherlands Royal Academy of Sciences. In 1959 he became honorary fellow of the British Institute of Physics and Physical Society and in 1960, Foreign Member of the Royal Society of Great Britain. In the same year he was elected to membership in the National Academy of Sciences of the United States and the American Academy of Arts and Sciences. In 1960 he became recipient of the F. London Prize (United States) and of the Max Planck Medal (West Germany). Lastly, in 1962 he was awarded the Nobel Prize in Physics "for his pioneering theories for condensed matter, especially liquid helium".

Landau's scientific influence was, of course, far from confined to his own disciples. He was deeply democratic in his life as a scientist (and in his life as a human being, for that matter; pomposity and deference to titles always remained foreign to him). Anyone, regardless of his scientific merits and title, could ask Landau for counsel and criticism (which were invariably precise and clear), on one condition only: the question must be businesslike instead of pertaining to what he detested most in science: empty philosophizing or vapidity and futility cloaked in pseudo-scientific sophistries. He had an acutely critical mind; this quality, along with his approach from the standpoint of profound physics, made discussion with him extremely attractive and useful.

In discussion he used to be ardent and incisive but not rude; witty and ironic but not caustic. The nameplate which he hung on the door of his office at the Ukrainian Physicotechnical Institute bore the inscription:

L. LANDAU
BEWARE, HE BITES!

With years his character and manner mellowed somewhat, but his enthusiasm for science and his uncompromising attitude toward science remained unchanged. And certainly his sharp exterior concealed a scientifically impartial attitude, a great heart and great kindness. However harsh and unsparing he may have been in his critical comments, he was just as intense in his desire to contribute with his advice to another man's success, and his approval, when he gave it, was just as ardent.

These traits of Landau's personality as a scientist and of his talent actually elevated him to the position of a supreme scientific judge, as it were, over his students and colleagues.† There is no doubt that this side of Landau's activities, his scientific and moral authority which exerted a restraining influence on frivolity in research, has also markedly contributed to the lofty level of our theoretical physics.

His constant scientific contact with a large number of students and colleagues also represented to Landau a source of knowledge. A unique aspect of his style of work was that, ever since long ago, since the Khar'kov years, he himself almost never read any scientific article or book but nevertheless he was always completely au courant with the latest news in physics.

† This position is symbolized in A. A. Yuzefovich's well-known friendly cartoon, "Dau said", reproduced here.

He derived this knowledge from numerous discussions and from the papers presented at the seminar held under his direction.

This seminar was held regularly once a week for nearly 30 years, and in the last years its sessions became gatherings of theoretical physicists from all Moscow. The presentation of papers at this seminar became a sacred duty for all students and co-workers, and Landau himself was extremely serious and thorough in selecting the material to be presented. He was interested and equally competent in every aspect of physics and the participants in the seminar did not find it easy to follow his train of thought in instantaneously switching from the discussion of, say, the properties of "strange" particles to the discussion of the energy spectrum of electrons in silicon. To Landau himself listening to the papers was never an empty formality: he did not rest until the essence of a study was completely elucidated and all traces of "philology" – unproved statements or propositions made on the principle of "why might it not" – therein were eliminated. As a result of such discussion and criticism many studies were condemned as "pathology" and Landau completely lost interest in them. On the other hand, articles that really contained new ideas or findings were included in the so-called "gold fund" and remained in Landau's memory for ever.

In fact, usually it was sufficient for him to know just the guiding idea of a study in order to reproduce all of its findings. As a rule, he found it easier to obtain them on his own than to follow in detail the author's reasoning. In this way he reproduced for himself and profoundly thought out most of the basic results obtained in all the domains of theoretical physics.† This probably also was the reason for his phenomenal ability to answer practically any question concerning physics that might be asked of him.

Landau's scientific style was free of the – unfortunately fairly widespread – tendency to complicate simple things (often on the grounds of generality and rigour which, however, usually turn out to be illusory). He himself always strove towards the opposite – to simplify complex things, to uncover in the most lucid manner the genuine simplicity of the laws underlying the natural phenomena. This ability of his, this skill at "trivializing" things as he himself used to say, was to him a matter of special pride.

The striving for simplicity and order was an inherent part of the structure of Landau's mind. It manifested itself not only in serious matters but also in semi-serious things as well as in his characteristic personal sense of humour.‡ Thus, he liked to classify everyone, from women according to the degree of their beauty, to theoretical physicists according to the signifi-

---

† Incidentally, this explains the absence of certain needed references in Landau's papers, which usually was not intentional. However, in some cases he could leave out a reference on purpose, if he considered the question too trivial; and he did have his own rather high standards on that matter.

‡ It is characteristic, however, that this trait was not a habit of Landau in his, so to speak, everyday outside life, in which he was not at all pedantically accurate and a "zone of disorder" would quite rapidly arise around him.

cance of their contribution to science. This last classification was based on a logarithmic scale of five: thus, a second-class physicist supposedly accomplished 10 times as much as a third-class physicist ("pathological types" were ranked in the fifth class). On this scale Einstein occupied the position $\frac{1}{2}$, while Bohr, Heisenberg, Schrödinger, Dirac and certain others were ranked in the first class. Landau modestly ranked himself for a long time in class $2\frac{1}{2}$ and it was only comparatively late in his life that he promoted himself to the second class.

He always worked hard (never at a desk, usually reclining on a divan at home). The recognition of the results of one's work is to a greater or lesser extent important to any scientist; it was, of course, also essential to Landau. But it can still be said that he attached much less importance to questions of priority than is ordinarily the case. And at any rate there is no doubt that his drive for work was inherently motivated not by desire for fame but by an inexhaustible curiosity and passion for exploring the laws of nature in their large and small manifestations. He never omitted a chance to repeat the elementary truth that one should never work for extraneous purposes, work merely for the sake of making a great discovery, for then nothing would be accomplished anyway.

The range of Landau's interests outside physics also was extremely wide. In addition to the exact sciences he loved history and was well-versed in it. He was also passionately interested in and deeply impressed by every genre of fine arts, though with the exception of music (and ballet).

Those who had the good fortune to be his students and friends for many years knew that our Dau, as his friends and comrades nicknamed him†, did not grow old. In his company boredom vanished. The brightness of his personality never grew dull and his scientific power remained strong. All the more senseless and frightful was the accident which put an end to his brilliant activity at its zenith.

*   *   *

Landau's articles, as a rule, display all the features of his characteristic scientific style: clarity and lucidity of physical statement of problems, the shortest and most elegant path towards their solution, no superfluities. Even now, after many years, the greater part of his articles does not require any revisions.

The brief review below is intended to provide only a tentative idea of the abundance and diversity of Landau's work and to clarify to some extent the place occupied by it in the history of physics, a place which may not always be obvious to the contemporary reader.

A characteristic feature of Landau's scientific creativity is its almost unprecedented breadth, which encompasses the whole of theoretical physics, from hydrodynamics to the quantum field theory. In our century, which is a century of increasingly narrow specialization, the scientific paths of his students also have been gradually diverging, but Landau himself unified them all, always retaining a truly astounding interest in everything. It may be that in him physics has lost one of the last great universalists.

Even a cursory examination of the bibliography of Landau's works shows that his life cannot be divided into any lengthy periods during which he worked only in some one domain of physics. Hence also the survey of his works is given not in chronological order but, insofar as possible, in thematic

---

† Landau himself liked to say that this name originated from the French spelling of his name: Landau = L'âne Dau (the ass Dau).

order. We shall begin with the works devoted to the general problems of quantum mechanics.

These include, in the first place, several of his early works. In the course of his studies of the radiation-damping problem he was the first to introduce the concept of incomplete quantum-mechanical description accomplished with the aid of quantities which were subsequently termed the density matrix [2]. In this article the density matrix was introduced in its energy representation.

Two articles [7, 9] are devoted to the calculation of the probabilities of quasiclassical processes. The difficulty of this problem stems from the fact that, by virtue of the exponential nature (with a large imaginary exponent) of the quasiclassical wave functions, the integrand in the matrix elements is a rapidly fluctuating quantity; this greatly complicates even an estimate of the integral; in fact, until Landau's work all studies of problems of this kind were erroneous. Landau was the first to provide a general method for the calculation of quasiclassical matrix elements and he also applied it to a number of specific processes.

In 1930 Landau (in collaboration with R. Peierls) published a detailed study of the limitations imposed by relativistic requirements on the quantum-mechanical description [6]; this article caused lively discussions at the time. Its basic result lies in determining the limits of the possibility of measuring the particle momentum within a finite time. This implied that in the relativistic quantum region it is not feasible to measure any dynamical variables characterizing the particles in their interaction, and that the only measurable quantities are the momenta (and polarizations) of free particles. Therein also lies the physical root of the difficulties that arise when methods of conventional quantum mechanics, employing concepts which become meaningless in the relativistic domain, are applied there. Landau returned to this problem in his last published article [100], in which he expressed his conviction that the ψ-operators, as carriers of unobservable information, and along with them the entire Hamiltonian method, should disappear from a future theory.

One of the reasons for this conviction was the results of the research into the foundations of quantum electrodynamics which Landau carried out during 1954–1955 (in collaboration with A. A. Abrikosov, I. M. Khalatnikov and I. Ya. Pomeranchuk) [78-81, 86]. These studies were based on the concept of the point interaction as the limit of "smeared" interaction when the smearing radius tends to zero. This made it possible to deal directly with finite expressions. Further, it proved possible to carry out the summation of the principal terms of the entire series of perturbation theory and this led to the derivation of asymptotic expressions (for the case of large momenta) for the fundamental quantities of quantum electrodynamics – the Green functions and the vertex part. These relations, in their own turn, were used to derive the relationship between the true charge and mass of the electron, on the one hand, and their "bare" values, on the other. Although these calculations proceeded on the premise of smallness of the "bare" charge, it was argued that the formula for the relation between true and bare charges retains its validity regardless of the magnitude of the bare charge. Then analysis of this formula shows that at the limit of point interaction the true charge becomes zero – the theory is "nullified".† (A review of the pertinent questions is provided in the articles [84, 89]).

Only the future will show the extent of the validity of the programme planned by Landau [100] for constructing a relativistic quantum field theory. He himself was energetically working in this direction during the last few years prior to his accident. As part of this programme, in particular, he had worked out a general method for determining the singularities of the quantities that occur in the diagram technique of quantum field theory [98].

In response to the discovery in 1956 of parity nonconservation in weak interactions, Landau immediately proposed the theory of a neutrino with fixed helicity ("two-component neutrino") [92]‡, and also suggested the principle of the conservation of "combined parity", as he termed the combined application of spatial inversion and charge conjugation. According to Landau, the symmetry of space would in this way be "saved" – the asymmetry is transferred to the particles themselves. This principle indeed proved to be more widely applicable than the law of parity conservation. As is known, however, in recent years processes not conserving combined parity have also been discovered; the meaning of this violation is at present still unclear.

A 1937 study [31] by Landau pertains to nuclear physics. This study represents a quantitative embodiment of the ideas proposed not long before by Bohr: the nucleus is examined by methods of statistical physics as a drop of "quantum fluid". It is noteworthy that this study did not make use of any far-reaching model conceptions, contrary to the previous practice of other investigators. In particular, the relationship between the mean distance between the levels of the compound nucleus and the width of the levels was established for the first time.

The absence of model conceptions is characteristic also of the theory of proton–proton scattering developed by Landau (in collaboration with Ya. A. Smorodinskii) [55]. The scattering cross-section in their study was expressed in terms of parameters whose meaning is not restricted by any specific assumptions concerning the particle interaction potential.

The study (in collaboration with Yu. B. Rumer) [36] of the cascade

---

† In connection with the search for a more rigorous proof of this statement, the article [100] contains the assertion, characteristic of Landau, that "the brevity of life does not allow us the luxury of spending time on problems which will lead to no new results".

‡ Simultaneously and independently, this theory was proposed by Salam and by Lee and Yang.

theory of electron showers in cosmic rays is an example of technical virtuosity; the physical foundations of this theory had been earlier formulated by a number of investigators, but a quantitative theory was essentially lacking. That study provided the mathematical apparatus which became the basis for all subsequent work in this domain. Landau himself took part in the further refinement of the shower theory by contributing two more articles, one on the particle angular distribution [43] and the other on secondary showers [44].

Of no smaller virtuosity was Landau's work dealing with the elaboration of Fermi's idea of the statistical nature of multiple particle production in collisions [74]. This study also represents a brilliant example of the methodological unity of theoretical physics in which the solution of a problem is accomplished by using the methods from a seemingly completely different domain. Landau showed that the process of multiple production includes the stage of the expansion of a "cloud" whose dimensions are large compared with the mean free path of particles in it; correspondingly, this stage should be described by equations of relativistic hydrodynamics. The solution of these equations required a number of ingenious techniques as well as a thorough analysis. Landau used to say that this study cost him more effort than any other problem that he had ever solved.

Landau always willingly responded to the requests and needs of the experimenters, e.g. by publishing the article [56] which established the energy distribution of the ionization losses of fast particles during passage through matter (previously only the theory of mean energy loss had existed).

Turning now to Landau's work on macroscopic physics, we begin with several articles representing his contribution to the physics of magnetism.

According to classical mechanics and statistics, a change in the pattern of movement of free electrons in a magnetic field cannot result in the appearance of new magnetic properties of the system. Landau was the first to elucidate the character of this motion in a magnetic field for the quantum case, and to show that quantization completely changes the situation, resulting in the appearance of diamagnetism of the free electron gas ("Landau diamagnetism" as this effect is now termed) [4]. The same study qualitatively predicted the periodic dependence of the magnetic susceptibility on the intensity of the magnetic field when this intensity is high. At the time (1930) this phenomenon had not yet been observed by anyone, and it was experimentally discovered only later (the De Haas–Van Alphen effect); a quantitative theory of this effect was presented by Landau in a later paper [38].

A short article published in 1933 [12] is of a significance greatly transcending the problem stated in its title – a possible explanation of the field dependence of the magnetic susceptibility of a particular class of substances at low temperatures. This article was the first to introduce the concept of antiferromagnetism (although it did not use this term) as a special phase of magnetic bodies differing in symmetry from the paramagnetic phase; accordingly, the transition from one state to the other, must occur at a rigorously definite point.† This article examined the particular model of a layered antiferromagnet with a strong ferromagnetic coupling in each layer and a weak antiferromagnetic coupling between the layers; a quantitative investigation of this case was carried out and the characteristic features of magnetic properties in the neighbourhood of the transition point were established. The method employed here by Landau was based on ideas which he subsequently elaborated in the general theory of second-order phase transitions.

Another paper concerns the theory of ferromagnetism. The idea of the structure of ferromagnetic bodies as consisting of elementary regions spontaneously magnetized in various directions ("magnetic domains," as the modern term goes) was expressed by P. Weiss as early as in 1907. However, there was no suitable approach to the question of the quantitative theory of this structure until Landau (in collaboration with E. M. Lifshitz) [18] showed in 1935 that this theory should be constructed on the basis of thermodynamic considerations and determined the form and dimensions of the domains for a typical case. The same study derived the macroscopic equation of the motion of the domain magnetization vector and, with its aid, developed the principles of the theory of the dispersion of the magnetic permeability of ferromagnets in an alternating magnetic field; in particular, it predicted the effect now known as ferromagnetic resonance.

A short communication published in 1933 [10] expressed the idea of the possibility of the "autolocalization" of an electron in a crystal lattice within the potential well produced by virtue of the polarization effect of the electron itself. This idea subsequently provided the basis for the so-called polaron theory of the conductivity of ionic crystals. Landau himself returned once more to these problems in a later study (in collaboration with S. I. Pekar) [67] dealing with the derivation of the equations of motion of the polaron in the external field.

Another short communication [14] reported on the results obtained by Landau (in collaboration with G. Placzek) concerning the structure of the Rayleigh scattering line in liquids or gases. As far back as the early 1920s Brillouin and Mandel'shtam showed that, owing to scattering by sound vibrations, this line must split into a doublet. Landau and Placzek drew attention to the attendant necessity of the existence of scattering by entropy

---

† Roughly a year earlier Néel (whose work was unknown to Landau) had predicted the possibility of existence of substances which, from the magnetic standpoint, consist of two sublattices with opposite moments. Néel, however, did not assume that a special state of matter is involved here, and instead he simply thought that a paramagnet with a positive exchange integral at low temperatures gradually turns into a structure consisting of several magnetic sublattices.

fluctuations, not accompanied by any change in frequency; as a result, a triplet should be observed instead of a doublet.†

Two of Landau's works pertain to plasma physics. One of these two [24] was the first to derive the transport equation with allowance for Coulomb interaction between particles; the slowness of decrease of these forces rendered inapplicable in this case the conventional methods for constructing transport equations. The other work [61], dealing with plasma oscillations, showed that, even under conditions when collisions between particles in the plasma can be disregarded, high-frequency oscillations will still attenuate ("Landau damping").‡

His work to compile one of the successive volumes of the *Course of Theoretical Physics* was to Landau a stimulus for a thorough study of hydrodynamics. Characteristically, he independently pondered and derived all the basic notions and results of this branch of science. His fresh and original perception led, in particular, to a new approach to the problem of the onset of turbulence and he elucidated the basic aspects of the process of the gradual development of unsteady flow with increase in the Reynolds number following the loss of stability by laminar motion and predicted qualitatively various alternatives possible in this case [52]. On investigating the qualitative properties of supersonic flow around bodies, he arrived at the unexpected discovery that in supersonic flow there must exist far from the body not one – as had been the conventional assumption – but two shock waves, one following the other [60]. Even in such a "classical" field as the jet theory he succeeded in finding a new and previously unnoticed exact solution for an axially symmetric "inundated" jet of a viscous incompressible fluid [51].

In Landau's scientific creative accomplishments an eminent position is occupied – both from the standpoint of direct significance and in terms of the consequent physical applications – by the theory of second-order phase transitions [29]; a first outline of the ideas underlying this theory is already contained in an earlier communication [17].‖ The concept of phase transitions of various orders had first been introduced by Ehrenfest in a purely formal manner, with respect to the order of the thermodynamic derivatives which could undergo a discontinuity at the transition point. The question of exactly which of these transitions can exist in reality, and what is their

---

† No detailed exposition of the conclusions and results of this study was ever published in article form. It is partly presented in the book by Landau and Lifshitz, *Electrodynamics of Continuous Media*, Pergamon, Oxford 1960, §96.

‡ It is interesting that this work was carried out by Landau as his response to the "philology" present, in his opinion, in previous studies dealing with this subject (e.g., the unjustified replacement of divergent integrals by their principal values). It was to prove his rightness that he occupied himself with this question.

‖ Landau himself applied this theory to the scattering of X-rays by crystals [32] and – in collaboration with I. M. Khalatnikov – to the absorption of sound in the neighbourhood of the transition point [82].

physical nature, had remained open, and previous interpretations had been fairly vague and unsubstantiated. Landau was the first to point to the profound connection between the possibility of existence of a continuous (in the sense of variation in the body's state) phase transition and the jump-like (discontinuous) change in some symmetry property of the body at the transition point. He also showed that far from just any change in symmetry is possible at that transition point and provided a method which makes it possible to determine the permissible types of change in symmetry. The quantitative theory developed by Landau was based on the assumption of the regularity of the expansion of thermodynamic quantities in the neighbourhood of the transition point. It is now clear that such a theory, which fails to allow for possible singularities of these quantities at the transition point, does not reflect all the properties of phase transitions. The question of the nature of these singularities was of great interest to Landau and during the last years of his activity he worked a great deal on this difficult problem without, however, succeeding in arriving at any definite conclusions.

The phenomenological theory of superconductivity developed in 1950 by Landau (in collaboration with V. L. Ginzburg) [73] also was constructed in the spirit of the theory of phase transitions; subsequently it became, in particular, the basis for the theory of superconducting alloys. This theory involves a number of variables and parameters whose meaning was not completely clear at the time it was originally developed and became understandable only after the appearance in 1957 of the microscopic theory of superconductivity, which made possible a rigorous substantiation of the Ginzburg–Landau equations and a determination of the region of their applicability. In this connection, the story (recounted by V. L. Ginzburg) of an erroneous statement contained in the original article by Landau and Ginzburg is instructive. The basic equation of the theory, defining the effective wave function Ψ of superconducting electrons, contains the field vector potential **A** in the term

$$\frac{1}{2m}\left(-i\hbar\,\nabla\,-\,\frac{e^*\mathbf{A}}{c}\right)\Psi,$$

which is completely analogous to the corresponding term in the Schrödinger equation. It might be thought that in the phenomenological theory the parameter e^* should represent some effective charge which does not have to be directly related to the charge of the free electron e. Landau, however, refuted this hypothesis by pointing out that the effective charge is not universal and would depend on various factors (pressure, composition of specimen, etc.); then in an inhomogeneous specimen the charge e^* would be a function of coordinates and this would disturb the gauge invariance of the theory. Hence the article stated that ". . . there is no reason to consider the charge e^* as different from the electronic charge". We now know that in reality e^* coincides with the charge of the Cooper electron pair, i.e.,

$e^* = 2e$ and not e. This value of e^* could, of course, have been predicted only on the basis of the idea of electron pairing which underlies the microscopic theory of superconductivity. But the value $2e$ is as universal as e and hence Landau's argument in itself was valid.

Another of Landau's contributions to the physics of superconductivity was to elucidate the nature of the so-called intermediate state. The concept of this state was first introduced by Peierls and F. London (1936) to account for the observed fact that the transition to the superconducting state in a magnetic field is gradual. Their theory was purely phenomenological, however, and the question of the nature of the intermediate state had remained open. Landau showed that this state is not a new state and that in reality a superconductor in that state consists of successive thin layers of normal and superconducting phases. In 1937 Landau [30] considered a model in which these layers emerge to the surface of the specimen; using an elegant and ingenious method he succeeded in completely determining the shape and dimensions of the layers in such a model.† In 1938 he proposed a new variant of the theory, according to which the layers repeatedly branch out on emerging to the surface; such a structure should be thermodynamically more favourable, given sufficiently large dimensions of the specimen.‡

But the most significant contribution that physics owes to Landau is his theory of quantum liquids. The significance of this new discipline at present is steadily growing; there is no doubt that its development in recent decades has produced a revolutionary effect on other domains of physics as well – on solid-state physics and even on nuclear physics.

The superfluidity theory was created by Landau during 1940–1941 soon after Kapitza's discovery towards the end of 1937 of this fundamental property of helium II. Prior to it, the premises for understanding the physical nature of the phase transition observed in liquid helium had been essentially lacking and it is not surprising that the previous interpretations of this phenomenon now seem even naive.‖ The completeness with which the theory of helium II had been constructed by Landau from the very beginning is remarkable: already his first classic paper [46] on this subject contained practically all the principal ideas of both the microscopic theory of helium II and the macroscopic theory constructed on its basis – the thermodynamics and hydrodynamics of this fluid.

Underlying Landau's theory is the concept of quasiparticles (elementary excitations) constituting the energy spectrum of helium II. Landau was in fact the first to pose the question of the energy spectrum of a macroscopic

---

† Landau himself wrote concerning this matter that "amazingly enough an exact determination of the shape of the layers proves to be possible" [30].

‡ A detailed description of this work was published in 1943 [49].

‖ Thus, Landau himself in his work on the theory of phase transitions [29] considered whether helium II is a liquid crystal, even though he emphasized the dubiousness of this assumption.

body in such a very general form, and it was he, too, who discovered the nature of the spectrum for a quantum fluid of the type to which liquid helium (He^4 isotope) belongs – or, as it is now termed, of the Bose type. In his 1941 work Landau assumed that the spectrum of elementary excitations consists of two branches: phonons, with a linear dependence of energy ε on momentum $\mathbf{p}$, and "rotons", with a quadratic dependence, separated from the ground state by an energy gap. Subsequently he found that such a form of spectrum is not satisfactory from the theoretical standpoint (as it would be unstable) and careful analysis of the more complete and exact experimental data that had by then become available led him in 1946 to establish the now famous spectrum containing only one branch in which the "rotons" correspond to a minimum on the curve of $\varepsilon(\mathbf{p})$. The macroscopic concepts of the theory of superfluidity are widely known. Basically they reduce to the idea of two motions simultaneously occurring in the fluid – "normal" motion and "superfluid" motion, which may be visualized as motions of two "fluid components".† Normal motion is accompanied by internal friction, as in conventional fluids. The determination of the viscosity coefficient represents a kinetic problem which requires an analysis of the processes of the onset of an equilibrium in the "gas of quasiparticles"; the principles of the theory of the viscosity of helium II were developed by Landau (in collaboration with I. M. Khalatnikov) in 1949 [69, 70]. Lastly, yet another investigation (carried out in collaboration with I. Ya. Pomeranchuk) [64] dealt with the problem of the behaviour of extraneous atoms in helium; it was shown, in particular, that any atom of this kind will become part of the "normal component" of the fluid regardless of whether the impurity substance itself does or does not display the property of superfluidity – contrary to the incorrect view previously held in the literature.

The liquid isotope He^3 is a quantum liquid of another type – the Fermi type as it is now termed. Although its properties are not as striking as the properties of liquid He^4, they are no less interesting from the standpoint of basic theory. A theory of liquids of this kind was developed by Landau and presented by him in three papers published during 1956–1958. The first two of these [90, 91] established the nature of the energy spectrum of Fermi liquids, considered their thermodynamic properties and established the kinetic equation for the relaxation processes occurring in these liquids. His study of the kinetic equation led Landau to predict a special type of vibra-

---

† Some of the ideas of the "two-component" macroscopic description of liquid helium were introduced independently of Landau by L. Tisza (although without providing a clear physical interpretation of them). His detailed article published in France in 1940 was, owing to wartime conditions, not received in the USSR until 1943 and the brief note of 1938 in the *Comptes rendus* of the Paris Académie des Sciences had unfortunately remained unnoticed. A criticism of the quantitative aspects of Tisza's theory was provided by Landau in the article [66].

tional process in liquid He^3 in the neighbourhood of absolute zero, which he termed zeroth sound. The third paper [95] presented a rigorous microscopic substantiation of the transport equation, whose earlier derivation had contained a number of intuitive assumptions.

Concluding this brief and far from complete survey, it only remains to be repeated that to physicists there is no need to emphasize the significance of Landau's contribution to theoretical physics. His accomplishments are of lasting significance and will for ever remain part of science.

CHAPTER I

THE EQUATIONS OF MOTION

§1. Generalised co-ordinates

ONE of the fundamental concepts of mechanics is that of a *particle*.† By this we mean a body whose dimensions may be neglected in describing its motion. The possibility of so doing depends, of course, on the conditions of the problem concerned. For example, the planets may be regarded as particles in considering their motion about the Sun, but not in considering their rotation about their axes.

The position of a particle in space is defined by its radius vector $\mathbf{r}$, whose components are its Cartesian co-ordinates x, y, z. The derivative $\mathbf{v} = d\mathbf{r}/dt$ of $\mathbf{r}$ with respect to the time t is called the *velocity* of the particle, and the second derivative $d^2\mathbf{r}/dt^2$ is its *acceleration*. In what follows we shall, as is customary, denote differentiation with respect to time by placing a dot above a letter: $\mathbf{v} = \dot{\mathbf{r}}$.

To define the position of a system of N particles in space, it is necessary to specify N radius vectors, i.e. $3N$ co-ordinates. The number of independent quantities which must be specified in order to define uniquely the position of any system is called the number of *degrees of freedom*; here, this number is $3N$. These quantities need not be the Cartesian co-ordinates of the particles, and the conditions of the problem may render some other choice of co-ordinates more convenient. Any s quantities q_1, q_2, ..., q_s which completely define the position of a system with s degrees of freedom are called *generalised co-ordinates* of the system, and the derivatives $\dot{q}_i$ are called its *generalised velocities*.

When the values of the generalised co-ordinates are specified, however, the "mechanical state" of the system at the instant considered is not yet determined in such a way that the position of the system at subsequent instants can be predicted. For given values of the co-ordinates, the system can have any velocities, and these affect the position of the system after an infinitesimal time interval dt.

If all the co-ordinates and velocities are simultaneously specified, it is known from experience that the state of the system is completely determined and that its subsequent motion can, in principle, be calculated. Mathematically, this means that, if all the co-ordinates q and velocities $\dot{q}$ are given at some instant, the accelerations $\ddot{q}$ at that instant are uniquely defined.‡

---

† Sometimes called in Russian a *material point*.
‡ For brevity, we shall often conventionally denote by q the set of all the co-ordinates $q_1, q_2, ..., q_s$, and similarly by $\dot{q}$ the set of all the velocities.

The relations between the accelerations, velocities and co-ordinates are called the *equations of motion*. They are second-order differential equations for the functions $q(t)$, and their integration makes possible, in principle, the determination of these functions and so of the path of the system.

§2. The principle of least action

The most general formulation of the law governing the motion of mechanical systems is the *principle of least action* or *Hamilton's principle*, according to which every mechanical system is characterised by a definite function $L(q_1, q_2, ..., q_s, \dot{q}_1, \dot{q}_2, ..., \dot{q}_s, t)$, or briefly $L(q, \dot{q}, t)$, and the motion of the system is such that a certain condition is satisfied.

Let the system occupy, at the instants t_1 and t_2, positions defined by two sets of values of the co-ordinates, $q^{(1)}$ and $q^{(2)}$. Then the condition is that the system moves between these positions in such a way that the integral

$$S = \int_{t_1}^{t_2} L(q, \dot{q}, t) \, dt \qquad (2.1)$$

takes the least possible value.† The function L is called the *Lagrangian* of the system concerned, and the integral (2.1) is called the *action*.

The fact that the Lagrangian contains only q and $\dot{q}$, but not the higher derivatives $\ddot{q}$, $\dddot{q}$, etc., expresses the result already mentioned, that the mechanical state of the system is completely defined when the co-ordinates and velocities are given.

Let us now derive the differential equations which solve the problem of minimising the integral (2.1). For simplicity, we shall at first assume that the system has only one degree of freedom, so that only one function $q(t)$ has to be determined.

Let $q = q(t)$ be the function for which S is a minimum. This means that S is increased when $q(t)$ is replaced by any function of the form

$$q(t) + \delta q(t), \qquad (2.2)$$

where $\delta q(t)$ is a function which is small everywhere in the interval of time from t_1 to t_2; $\delta q(t)$ is called a *variation* of the function $q(t)$. Since, for $t = t_1$ and for $t = t_2$, all the functions (2.2) must take the values $q^{(1)}$ and $q^{(2)}$ respectively, it follows that

$$\delta q(t_1) = \delta q(t_2) = 0. \qquad (2.3)$$

---

† It should be mentioned that this formulation of the principle of least action is not always valid for the entire path of the system, but only for any sufficiently short segment of the path. The integral (2.1) for the entire path must have an extremum, but not necessarily a minimum. This fact, however, is of no importance as regards the derivation of the equations of motion, since only the extremum condition is used.

The change in S when q is replaced by $q + \delta q$ is

$$\int_{t_1}^{t_2} L(q + \delta q, \dot{q} + \delta \dot{q}, t)\, dt - \int_{t_1}^{t_2} L(q, \dot{q}, t)\, dt.$$

When this difference is expanded in powers of δq and $\delta \dot{q}$ in the integrand, the leading terms are of the first order. The necessary condition for S to have a minimum† is that these terms (called the *first variation*, or simply the *variation*, of the integral) should be zero. Thus the principle of least action may be written in the form

$$\delta S = \delta \int_{t_1}^{t_2} L(q, \dot{q}, t)\, dt = 0, \tag{2.4}$$

or, effecting the variation,

$$\int_{t_1}^{t_2} \left(\frac{\partial L}{\partial q} \delta q + \frac{\partial L}{\partial \dot{q}} \delta \dot{q} \right) dt = 0.$$

Since $\delta \dot{q} = d\delta q/dt$, we obtain, on integrating the second term by parts,

$$\delta S = \left[\frac{\partial L}{\partial \dot{q}} \delta q \right]_{t_1}^{t_2} + \int_{t_1}^{t_2} \left(\frac{\partial L}{\partial q} - \frac{d}{dt} \frac{\partial L}{\partial \dot{q}} \right) \delta q\, dt = 0. \tag{2.5}$$

The conditions (2.3) show that the integrated term in (2.5) is zero. There remains an integral which must vanish for all values of δq. This can be so only if the integrand is zero identically. Thus we have

$$\frac{d}{dt} \left(\frac{\partial L}{\partial \dot{q}} \right) - \frac{\partial L}{\partial q} = 0.$$

When the system has more than one degree of freedom, the s different functions $q_i(t)$ must be varied independently in the principle of least action. We then evidently obtain s equations of the form

$$\frac{d}{dt} \left(\frac{\partial L}{\partial \dot{q}_i} \right) - \frac{\partial L}{\partial q_i} = 0 \qquad (i = 1, 2, ..., s). \tag{2.6}$$

These are the required differential equations, called in mechanics *Lagrange's equations*.‡ If the Lagrangian of a given mechanical system is known, the equations (2.6) give the relations between accelerations, velocities and co-ordinates, i.e. they are the equations of motion of the system.

---

† Or, in general, an extremum.
‡ In the calculus of variations they are Euler's equations for the formal problem of determining the extrema of an integral of the form (2.1).

Mathematically, the equations (2.6) constitute a set of s second-order equations for s unknown functions $q_i(t)$. The general solution contains $2s$ arbitrary constants. To determine these constants and thereby to define uniquely the motion of the system, it is necessary to know the initial conditions which specify the state of the system at some given instant, for example the initial values of all the co-ordinates and velocities.

Let a mechanical system consist of two parts A and B which would, if closed, have Lagrangians L_A and L_B respectively. Then, in the limit where the distance between the parts becomes so large that the interaction between them may be neglected, the Lagrangian of the whole system tends to the value

$$\lim L = L_A + L_B. \tag{2.7}$$

This additivity of the Lagrangian expresses the fact that the equations of motion of either of the two non-interacting parts cannot involve quantities pertaining to the other part.

It is evident that the multiplication of the Lagrangian of a mechanical system by an arbitrary constant has no effect on the equations of motion. From this, it might seem, the following important property of arbitrariness can be deduced: the Lagrangians of different isolated mechanical systems may be multiplied by different arbitrary constants. The additive property, however, removes this indefiniteness, since it admits only the simultaneous multiplication of the Lagrangians of all the systems by the same constant. This corresponds to the natural arbitrariness in the choice of the unit of measurement of the Lagrangian, a matter to which we shall return in §4.

One further general remark should be made. Let us consider two functions $L'(q, \dot{q}, t)$ and $L(q, \dot{q}, t)$, differing by the total derivative with respect to time of some function $f(q, t)$ of co-ordinates and time:

$$L'(q, \dot{q}, t) = L(q, \dot{q}, t) + \frac{\mathrm{d}}{\mathrm{d}t} f(q, t). \tag{2.8}$$

The integrals (2.1) calculated from these two functions are such that

$$S' = \int_{t_1}^{t_2} L'(q, \dot{q}, t)\,\mathrm{d}t = \int_{t_1}^{t_2} L(q, \dot{q}, t)\,\mathrm{d}t + \int_{t_1}^{t_2} \frac{\mathrm{d}f}{\mathrm{d}t}\,\mathrm{d}t = S + f(q^{(2)}, t_2) - f(q^{(1)}, t_1),$$

i.e. they differ by a quantity which gives zero on variation, so that the conditions $\delta S' = 0$ and $\delta S = 0$ are equivalent, and the form of the equations of motion is unchanged. Thus the Lagrangian is defined only to within an additive total time derivative of any function of co-ordinates and time.

§3. Galileo's relativity principle

In order to consider mechanical phenomena it is necessary to choose a *frame of reference*. The laws of motion are in general different in form for different frames of reference. When an arbitrary frame of reference is chosen, it may happen that the laws governing even very simple phenomena become very complex. The problem naturally arises of finding a frame of reference in which the laws of mechanics take their simplest form.

If we were to choose an arbitrary frame of reference, space would be inhomogeneous and anisotropic. This means that, even if a body interacted with no other bodies, its various positions in space and its different orientations would not be mechanically equivalent. The same would in general be true of time, which would likewise be inhomogeneous; that is, different instants would not be equivalent. Such properties of space and time would evidently complicate the description of mechanical phenomena. For example, a free body (i.e. one subject to no external action) could not remain at rest: if its velocity were zero at some instant, it would begin to move in some direction at the next instant.

It is found, however, that a frame of reference can always be chosen in which space is homogeneous and isotropic and time is homogeneous. This is called an *inertial frame*. In particular, in such a frame a free body which is at rest at some instant remains always at rest.

We can now draw some immediate inferences concerning the form of the Lagrangian of a particle, moving freely, in an inertial frame of reference. The homogeneity of space and time implies that the Lagrangian cannot contain explicitly either the radius vector $\mathbf{r}$ of the particle or the time t, i.e. L must be a function of the velocity $\mathbf{v}$ only. Since space is isotropic, the Lagrangian must also be independent of the direction of $\mathbf{v}$, and is therefore a function only of its magnitude, i.e. of $\mathbf{v}^2 = v^2$:

$$L = L(v^2). \tag{3.1}$$

Since the Lagrangian is independent of $\mathbf{r}$, we have $\partial L/\partial \mathbf{r} = 0$, and so Lagrange's equation is†

$$\frac{\mathrm{d}}{\mathrm{d}t}\left(\frac{\partial L}{\partial \mathbf{v}}\right) = 0,$$

whence $\partial L/\partial \mathbf{v} = \text{constant}$. Since $\partial L/\partial \mathbf{v}$ is a function of the velocity only, it follows that

$$\mathbf{v} = \text{constant}. \tag{3.2}$$

Thus we conclude that, in an inertial frame, any free motion takes place with a velocity which is constant in both magnitude and direction. This is the *law of inertia*.

If we consider, besides the inertial frame, another frame moving uniformly in a straight line relative to the inertial frame, then the laws of free motion in

---

† The derivative of a scalar quantity with respect to a vector is defined as the vector whose components are equal to the derivatives of the scalar with respect to the corresponding components of the vector.

the other frame will be the same as in the original frame: free motion takes place with a constant velocity.

Experiment shows that not only are the laws of free motion the same in the two frames, but the frames are entirely equivalent in all mechanical respects. Thus there is not one but an infinity of inertial frames moving, relative to one another, uniformly in a straight line. In all these frames the properties of space and time are the same, and the laws of mechanics are the same. This constitutes *Galileo's relativity principle*, one of the most important principles of mechanics.

The above discussion indicates quite clearly that inertial frames of reference have special properties, by virtue of which they should, as a rule, be used in the study of mechanical phenomena. In what follows, unless the contrary is specifically stated, we shall consider only inertial frames.

The complete mechanical equivalence of the infinity of such frames shows also that there is no "absolute" frame of reference which should be preferred to other frames.

The co-ordinates **r** and **r'** of a given point in two different frames of reference K and K', of which the latter moves relative to the former with velocity **V**, are related by

$$\mathbf{r} = \mathbf{r'} + \mathbf{V}t. \tag{3.3}$$

Here it is understood that time is the same in the two frames:

$$t = t'. \tag{3.4}$$

The assumption that time is absolute is one of the foundations of classical mechanics.†

Formulae (3.3) and (3.4) are called a *Galilean transformation*. Galileo's relativity principle can be formulated as asserting the invariance of the mechanical equations of motion under any such transformation.

§4. The Lagrangian for a free particle

Let us now go on to determine the form of the Lagrangian, and consider first of all the simplest case, that of the free motion of a particle relative to an inertial frame of reference. As we have already seen, the Lagrangian in this case can depend only on the square of the velocity. To discover the form of this dependence, we make use of Galileo's relativity principle. If an inertial frame K is moving with an infinitesimal velocity **ε** relative to another inertial frame K', then $\mathbf{v'} = \mathbf{v} + \mathbf{ε}$. Since the equations of motion must have the same form in every frame, the Lagrangian $L(v^2)$ must be converted by this transformation into a function L' which differs from $L(v^2)$, if at all, only by the total time derivative of a function of co-ordinates and time (see the end of §2).

---

† This assumption does not hold good in relativistic mechanics.

We have $L' = L(v'^2) = L(v^2 + 2\mathbf{v} \cdot \boldsymbol{\epsilon} + \epsilon^2)$. Expanding this expression in powers of $\boldsymbol{\epsilon}$ and neglecting terms above the first order, we obtain

$$L(v'^2) = L(v^2) + \frac{\partial L}{\partial v^2} 2\mathbf{v} \cdot \boldsymbol{\epsilon}.$$

The second term on the right of this equation is a total time derivative only if it is a linear function of the velocity $\mathbf{v}$. Hence $\partial L/\partial v^2$ is independent of the velocity, i.e. the Lagrangian is in this case proportional to the square of the velocity, and we write it as

$$L = \tfrac{1}{2}mv^2. \tag{4.1}$$

From the fact that a Lagrangian of this form satisfies Galileo's relativity principle for an infinitesimal relative velocity, it follows at once that the Lagrangian is invariant for a finite relative velocity $\mathbf{V}$ of the frames K and K'. For

$$L' = \tfrac{1}{2}mv'^2 = \tfrac{1}{2}m(\mathbf{v} + \mathbf{V})^2 = \tfrac{1}{2}mv^2 + m\mathbf{v} \cdot \mathbf{V} + \tfrac{1}{2}mV^2,$$

or

$$L' = L + \mathrm{d}(m\mathbf{r} \cdot \mathbf{V} + \tfrac{1}{2}mV^2 t)/\mathrm{d}t.$$

The second term is a total time derivative and may be omitted.

The quantity m which appears in the Lagrangian (4.1) for a freely moving particle is called the *mass* of the particle. The additive property of the Lagrangian shows that for a system of particles which do not interact we have[†]

$$L = \sum \tfrac{1}{2}m_a v_a^2. \tag{4.2}$$

It should be emphasised that the above definition of mass becomes meaningful only when the additive property is taken into account. As has been mentioned in §2, the Lagrangian can always be multiplied by any constant without affecting the equations of motion. As regards the function (4.2), such multiplication amounts to a change in the unit of mass; the ratios of the masses of different particles remain unchanged thereby, and it is only these ratios which are physically meaningful.

It is easy to see that the mass of a particle cannot be negative. For, according to the principle of least action, the integral

$$S = \int_1^2 \tfrac{1}{2}mv^2 \, \mathrm{d}t$$

has a minimum for the actual motion of the particle in space from point 1 to point 2. If the mass were negative, the action integral would take arbitrarily large negative values for a motion in which the particle rapidly left point 1 and rapidly approached point 2, and there would be no minimum.[‡]

[†] We shall use the suffixes $a, b, c, \ldots$ to distinguish the various particles, and $i, k, l, \ldots$ to distinguish the co-ordinates.

[‡] The argument is not affected by the point mentioned in the first footnote to §2; for $m < 0$, the integral could not have a minimum even for a short segment of the path.

It is useful to notice that

$$v^2 = (\mathrm{d}l/\mathrm{d}t)^2 = (\mathrm{d}l)^2/(\mathrm{d}t)^2. \qquad (4.3)$$

Hence, to obtain the Lagrangian, it is sufficient to find the square of the element of arc $\mathrm{d}l$ in a given system of co-ordinates. In Cartesian co-ordinates, for example, $\mathrm{d}l^2 = \mathrm{d}x^2 + \mathrm{d}y^2 + \mathrm{d}z^2$, and so

$$L = \tfrac{1}{2}m(\dot{x}^2 + \dot{y}^2 + \dot{z}^2). \qquad (4.4)$$

In cylindrical co-ordinates $\mathrm{d}l^2 = \mathrm{d}r^2 + r^2\,\mathrm{d}\phi^2 + \mathrm{d}z^2$, whence

$$L = \tfrac{1}{2}m(\dot{r}^2 + r^2\dot{\phi}^2 + \dot{z}^2). \qquad (4.5)$$

In spherical co-ordinates $\mathrm{d}l^2 = \mathrm{d}r^2 + r^2\,\mathrm{d}\theta^2 + r^2\sin^2\theta\,\mathrm{d}\phi^2$, and

$$L = \tfrac{1}{2}m(\dot{r}^2 + r^2\dot{\theta}^2 + r^2\dot{\phi}^2\sin^2\theta). \qquad (4.6)$$

§5. The Lagrangian for a system of particles

Let us now consider a system of particles which interact with one another but with no other bodies. This is called a *closed system*. It is found that the interaction between the particles can be described by adding to the Lagrangian (4.2) for non-interacting particles a certain function of the co-ordinates, which depends on the nature of the interaction.† Denoting this function by $-U$, we have

$$L = \sum \tfrac{1}{2}m_a v_a{}^2 - U(\mathbf{r}_1, \mathbf{r}_2, ...), \qquad (5.1)$$

where $\mathbf{r}_a$ is the radius vector of the ath particle. This is the general form of the Lagrangian for a closed system. The sum $T = \sum \tfrac{1}{2}m_a v_a{}^2$ is called the *kinetic energy*, and U the *potential energy*, of the system. The significance of these names is explained in §6.

The fact that the potential energy depends only on the positions of the particles at a given instant shows that a change in the position of any particle instantaneously affects all the other particles. We may say that the interactions are instantaneously propagated. The necessity for interactions in classical mechanics to be of this type is closely related to the premises upon which the subject is based, namely the absolute nature of time and Galileo's relativity principle. If the propagation of interactions were not instantaneous, but took place with a finite velocity, then that velocity would be different in different frames of reference in relative motion, since the absoluteness of time necessarily implies that the ordinary law of composition of velocities is applicable to all phenomena. The laws of motion for interacting bodies would then be different in different inertial frames, a result which would contradict the relativity principle.

In §3 only the homogeneity of time has been spoken of. The form of the Lagrangian (5.1) shows that time is both homogeneous and isotropic, i.e. its

---

† This statement is valid in classical mechanics. Relativistic mechanics is not considered in this book.

properties are the same in both directions. For, if t is replaced by $-t$, the Lagrangian is unchanged, and therefore so are the equations of motion. In other words, if a given motion is possible in a system, then so is the reverse motion (that is, the motion in which the system passes through the same states in the reverse order). In this sense all motions which obey the laws of classical mechanics are reversible.

Knowing the Lagrangian, we can derive the equations of motion:

$$\frac{d}{dt}\frac{\partial L}{\partial \mathbf{v}_a} = \frac{\partial L}{\partial \mathbf{r}_a}. \tag{5.2}$$

Substitution of (5.1) gives

$$m_a\, d\mathbf{v}_a/dt = -\partial U/\partial \mathbf{r}_a. \tag{5.3}$$

In this form the equations of motion are called *Newton's equations* and form the basis of the mechanics of a system of interacting particles. The vector

$$\mathbf{F} = -\partial U/\partial \mathbf{r}_a \tag{5.4}$$

which appears on the right-hand side of equation (5.3) is called the *force* on the ath particle. Like U, it depends only on the co-ordinates of the particles, and not on their velocities. The equation (5.3) therefore shows that the acceleration vectors of the particles are likewise functions of their co-ordinates only.

The potential energy is defined only to within an additive constant, which has no effect on the equations of motion. This is a particular case of the non-uniqueness of the Lagrangian discussed at the end of §2. The most natural and most usual way of choosing this constant is such that the potential energy tends to zero as the distances between the particles tend to infinity.

If we use, to describe the motion, arbitrary generalised co-ordinates q_i instead of Cartesian co-ordinates, the following transformation is needed to obtain the new Lagrangian:

$$x_a = f_a(q_1, q_2, ..., q_s), \quad \dot{x}_a = \sum_k \frac{\partial f_a}{\partial q_k}\dot{q}_k, \text{ etc.}$$

Substituting these expressions in the function $L = \frac{1}{2}\Sigma m_a(\dot{x}_a{}^2 + \dot{y}_a{}^2 + \dot{z}_a{}^2) - U$, we obtain the required Lagrangian in the form

$$L = \frac{1}{2}\sum_{i,k} a_{ik}(q)\dot{q}_i\dot{q}_k - U(q), \tag{5.5}$$

where the a_{ik} are functions of the co-ordinates only. The kinetic energy in generalised co-ordinates is still a quadratic function of the velocities, but it may depend on the co-ordinates also.

Hitherto we have spoken only of closed systems. Let us now consider a system A which is not closed and interacts with another system B executing a given motion. In such a case we say that the system A moves in a given external field (due to the system B). Since the equations of motion are obtained from the principle of least action by independently varying each of the co-ordinates (i.e. by proceeding as if the remainder were given quantities), we can find the Lagrangian L_A of the system A by using the Lagrangian L of the whole system $A+B$ and replacing the co-ordinates q_B therein by given functions of time.

Assuming that the system $A+B$ is closed, we have $L = T_A(q_A, \dot{q}_A) + T_B(q_B, \dot{q}_B) - U(q_A, q_B)$, where the first two terms are the kinetic energies of the systems A and B and the third term is their combined potential energy. Substituting for q_B the given functions of time and omitting the term $T[q_B(t), \dot{q}_B(t)]$ which depends on time only, and is therefore the total time derivative of a function of time, we obtain $L_A = T_A(q_A, \dot{q}_A) - U[q_A, q_B(t)]$. Thus the motion of a system in an external field is described by a Lagrangian of the usual type, the only difference being that the potential energy may depend explicitly on time.

For example, when a single particle moves in an external field, the general form of the Lagrangian is

$$L = \tfrac{1}{2}mv^2 - U(\mathbf{r}, t),\tag{5.6}$$

and the equation of motion is

$$m\,\dot{\mathbf{v}} = -\partial U/\partial \mathbf{r}.\tag{5.7}$$

A field such that the same force $\mathbf{F}$ acts on a particle at any point in the field is said to be *uniform*. The potential energy in such a field is evidently

$$U = -\mathbf{F}\cdot\mathbf{r}.\tag{5.8}$$

To conclude this section, we may make the following remarks concerning the application of Lagrange's equations to various problems. It is often necessary to deal with mechanical systems in which the interaction between different bodies (or particles) takes the form of *constraints*, i.e. restrictions on their relative position. In practice, such constraints are effected by means of rods, strings, hinges and so on. This introduces a new factor into the problem, in that the motion of the bodies results in friction at their points of contact, and the problem in general ceases to be one of pure mechanics (see §25). In many cases, however, the friction in the system is so slight that its effect on the motion is entirely negligible. If the masses of the constraining elements of the system are also negligible, the effect of the constraints is simply to reduce the number of degrees of freedom s of the system to a value less than $3N$. To determine the motion of the system, the Lagrangian (5.5) can again be used, with a set of independent generalised co-ordinates equal in number to the actual degrees of freedom.

PROBLEMS

Find the Lagrangian for each of the following systems when placed in a uniform gravitational field (acceleration g).

PROBLEM 1. A coplanar double pendulum (Fig. 1).

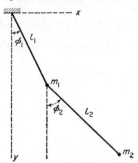

FIG. 1

SOLUTION. We take as co-ordinates the angles ϕ_1 and ϕ_2 which the strings l_1 and l_2 make with the vertical. Then we have, for the particle m_1, $T_1 = \frac{1}{2}m_1 l_1^2 \dot{\phi}_1^2$, $U = -m_1 g l_1 \cos \phi_1$. In order to find the kinetic energy of the second particle, we express its Cartesian co-ordinates x_2, y_2 (with the origin at the point of support and the y-axis vertically downwards) in terms of the angles ϕ_1 and ϕ_2: $x_2 = l_1 \sin \phi_1 + l_2 \sin \phi_2$, $y_2 = l_1 \cos \phi_1 + l_2 \cos \phi_2$. Then we find

$$T_2 = \frac{1}{2}m_2(\dot{x}_2^2 + \dot{y}_2^2)$$

$$= \frac{1}{2}m_2[l_1^2\dot{\phi}_1^2 + l_2^2\dot{\phi}_2^2 + 2l_1 l_2 \cos(\phi_1 - \phi_2)\dot{\phi}_1\dot{\phi}_2].$$

Finally

$$L = \frac{1}{2}(m_1 + m_2)l_1^2\dot{\phi}_1^2 + \frac{1}{2}m_2 l_2^2\dot{\phi}_2^2 + m_2 l_1 l_2 \dot{\phi}_1\dot{\phi}_2 \cos(\phi_1 - \phi_2) + (m_1 + m_2)g l_1 \cos \phi_1 + m_2 g l_2 \cos \phi_2.$$

PROBLEM 2. A simple pendulum of mass m_2, with a mass m_1 at the point of support which can move on a horizontal line lying in the plane in which m_2 moves (Fig. 2).

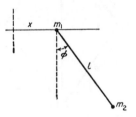

FIG. 2

SOLUTION. Using the co-ordinate x of m_1 and the angle ϕ between the string and the vertical, we have

$$L = \frac{1}{2}(m_1 + m_2)\dot{x}^2 + \frac{1}{2}m_2(l^2\dot{\phi}^2 + 2l\dot{x}\dot{\phi} \cos \phi) + m_2 g l \cos \phi.$$

PROBLEM 3. A simple pendulum of mass m whose point of support (a) moves uniformly on a vertical circle with constant frequency γ (Fig. 3), (b) oscillates horizontally in the plane of motion of the pendulum according to the law $x = a \cos \gamma t$, (c) oscillates vertically according to the law $y = a \cos \gamma t$.

SOLUTION. (a) The co-ordinates of m are $x = a \cos \gamma t + l \sin \phi$, $y = -a \sin \gamma t + l \cos \phi$. The Lagrangian is

$$L = \frac{1}{2}ml^2\dot{\phi}^2 + ml a \gamma^2 \sin(\phi - \gamma t) + mgl \cos \phi;$$

here terms depending only on time have been omitted, together with the total time derivative of $ml a \gamma \cos(\phi - \gamma t)$.

(b) The co-ordinates of m are $x = a \cos \gamma t + l \sin \phi$, $y = l \cos \phi$. The Lagrangian is (omitting total derivatives)

$$L = \tfrac{1}{2}ml^2\dot\phi^2 + mla\gamma^2 \cos \gamma t \sin \phi + mgl \cos \phi.$$

(c) Similarly

$$L = \tfrac{1}{2}ml^2\dot\phi^2 + mla\gamma^2 \cos \gamma t \cos \phi + mgl \cos \phi.$$

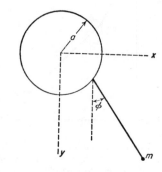

FIG. 3

PROBLEM 4. The system shown in Fig. 4. The particle m_2 moves on a vertical axis and the whole system rotates about this axis with a constant angular velocity Ω.

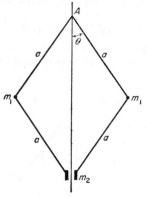

FIG. 4

SOLUTION. Let θ be the angle between one of the segments a and the vertical, and ϕ the angle of rotation of the system about the axis; $\dot\phi = \Omega$. For each particle m_1, the infinitesimal displacement is given by $dl_1^2 = a^2\, d\theta^2 + a^2 \sin^2 \theta\, d\phi^2$. The distance of m_2 from the point of support A is $2a \cos \theta$, and so $dl_2 = -2a \sin \theta\, d\theta$. The Lagrangian is

$$L = m_1a^2(\dot\theta^2 + \Omega^2 \sin^2\theta) + 2m_2a^2\dot\theta^2 \sin^2\theta + 2(m_1+m_2)ga \cos \theta.$$

CHAPTER II

CONSERVATION LAWS

§6. Energy

DURING the motion of a mechanical system, the $2s$ quantities q_i and $\dot{q}_i$ ($i = 1, 2, \ldots, s$) which specify the state of the system vary with time. There exist, however, functions of these quantities whose values remain constant during the motion, and depend only on the initial conditions. Such functions are called *integrals of the motion*.

The number of independent integrals of the motion for a closed mechanical system with s degrees of freedom is $2s-1$. This is evident from the following simple arguments. The general solution of the equations of motion contains $2s$ arbitrary constants (see the discussion following equation (2.6)). Since the equations of motion for a closed system do not involve the time explicitly, the choice of the origin of time is entirely arbitrary, and one of the arbitrary constants in the solution of the equations can always be taken as an additive constant t_0 in the time. Eliminating $t+t_0$ from the $2s$ functions $q_i = q_i(t+t_0, C_1, C_2, \ldots, C_{2s-1})$, $\dot{q}_i = \dot{q}_i(t+t_0, C_1, C_2, \ldots, C_{2s-1})$, we can express the $2s-1$ arbitrary constants $C_1, C_2, \ldots, C_{2s-1}$ as functions of q and $\dot{q}$, and these functions will be integrals of the motion.

Not all integrals of the motion, however, are of equal importance in mechanics. There are some whose constancy is of profound significance, deriving from the fundamental homogeneity and isotropy of space and time. The quantities represented by such integrals of the motion are said to be *conserved*, and have an important common property of being additive: their values for a system composed of several parts whose interaction is negligible are equal to the sums of their values for the individual parts.

It is to this additivity that the quantities concerned owe their especial importance in mechanics. Let us suppose, for example, that two bodies interact during a certain interval of time. Since each of the additive integrals of the whole system is, both before and after the interaction, equal to the sum of its values for the two bodies separately, the conservation laws for these quantities immediately make possible various conclusions regarding the state of the bodies after the interaction, if their states before the interaction are known.

Let us consider first the conservation law resulting from the *homogeneity of time*. By virtue of this homogeneity, the Lagrangian of a closed system does not depend explicitly on time. The total time derivative of the Lagrangian can therefore be written

$$\frac{dL}{dt} = \sum_i \frac{\partial L}{\partial q_i}\dot{q}_i + \sum_i \frac{\partial L}{\partial \dot{q}_i}\ddot{q}_i.$$

If L depended explicitly on time, a term $\partial L/\partial t$ would have to be added on the right-hand side. Replacing $\partial L/\partial q_i$, in accordance with Lagrange's equations, by $(\mathrm{d}/\mathrm{d}t)\,\partial L/\partial \dot{q}_i$, we obtain

$$\frac{\mathrm{d}L}{\mathrm{d}t} = \sum_i \dot{q}_i \frac{\mathrm{d}}{\mathrm{d}t}\left(\frac{\partial L}{\partial \dot{q}_i}\right) + \sum_i \frac{\partial L}{\partial \dot{q}_i}\ddot{q}_i$$

$$= \sum_i \frac{\mathrm{d}}{\mathrm{d}t}\left(\dot{q}_i \frac{\partial L}{\partial \dot{q}_i}\right)$$

or

$$\frac{\mathrm{d}}{\mathrm{d}t}\left(\sum_i \dot{q}_i \frac{\partial L}{\partial \dot{q}_i} - L\right) = 0.$$

Hence we see that the quantity

$$E \equiv \sum_i \dot{q}_i \frac{\partial L}{\partial \dot{q}_i} - L \tag{6.1}$$

remains constant during the motion of a closed system, i.e. it is an integral of the motion; it is called the *energy* of the system. The additivity of the energy follows immediately from that of the Lagrangian, since (6.1) shows that it is a linear function of the latter.

The law of conservation of energy is valid not only for closed systems, but also for those in a constant external field (i.e. one independent of time): the only property of the Lagrangian used in the above derivation, namely that it does not involve the time explicitly, is still valid. Mechanical systems whose energy is conserved are sometimes called *conservative* systems.

As we have seen in §5, the Lagrangian of a closed system (or one in a constant field) is of the form $L = T(q, \dot{q}) - U(q)$, where T is a quadratic function of the velocities. Using Euler's theorem on homogeneous functions, we have

$$\sum_i \dot{q}_i \frac{\partial L}{\partial \dot{q}_i} = \sum_i \dot{q}_i \frac{\partial T}{\partial \dot{q}_i} = 2T.$$

Substituting this in (6.1) gives

$$E = T(q, \dot{q}) + U(q); \tag{6.2}$$

in Cartesian co-ordinates,

$$E = \sum_a \tfrac{1}{2}m_a v_a^2 + U(\mathbf{r}_1, \mathbf{r}_2, \ldots). \tag{6.3}$$

Thus the energy of the system can be written as the sum of two quite different terms: the kinetic energy, which depends on the velocities, and the potential energy, which depends only on the co-ordinates of the particles.

§7. Momentum

A second conservation law follows from the *homogeneity of space*. By virtue of this homogeneity, the mechanical properties of a closed system are unchanged by any parallel displacement of the entire system in space. Let us therefore consider an infinitesimal displacement ϵ, and obtain the condition for the Lagrangian to remain unchanged.

A parallel displacement is a transformation in which every particle in the system is moved by the same amount, the radius vector $\mathbf{r}$ becoming $\mathbf{r}+\epsilon$. The change in L resulting from an infinitesimal change in the co-ordinates, the velocities of the particles remaining fixed, is

$$\delta L = \sum_a \frac{\partial L}{\partial \mathbf{r}_a} \cdot \delta \mathbf{r}_a = \epsilon \cdot \sum_a \frac{\partial L}{\partial \mathbf{r}_a},$$

where the summation is over the particles in the system. Since ϵ is arbitrary, the condition $\delta L = 0$ is equivalent to

$$\sum_a \partial L/\partial \mathbf{r}_a = 0. \tag{7.1}$$

From Lagrange's equations (5.2) we therefore have

$$\sum_a \frac{d}{dt} \frac{\partial L}{\partial \mathbf{v}_a} = \frac{d}{dt} \sum_a \frac{\partial L}{\partial \mathbf{v}_a} = 0.$$

Thus we conclude that, in a closed mechanical system, the vector

$$\mathbf{P} \equiv \sum_a \partial L/\partial \mathbf{v}_a \tag{7.2}$$

remains constant during the motion; it is called the *momentum* of the system. Differentiating the Lagrangian (5.1), we find that the momentum is given in terms of the velocities of the particles by

$$\mathbf{P} = \sum_a m_a \mathbf{v}_a. \tag{7.3}$$

The additivity of the momentum is evident. Moreover, unlike the energy, the momentum of the system is equal to the sum of its values $\mathbf{p}_a = m_a \mathbf{v}_a$ for the individual particles, whether or not the interaction between them can be neglected.

The three components of the momentum vector are all conserved only in the absence of an external field. The individual components may be conserved even in the presence of a field, however, if the potential energy in the field does not depend on all the Cartesian co-ordinates. The mechanical properties of the system are evidently unchanged by a displacement along the axis of a co-ordinate which does not appear in the potential energy, and so the corresponding component of the momentum is conserved. For example, in a uniform field in the z-direction, the x and y components of momentum are conserved.

The equation (7.1) has a simple physical meaning. The derivative $\partial L/\partial \mathbf{r}_a = -\partial U/\partial \mathbf{r}_a$ is the force $\mathbf{F}_a$ acting on the ath particle. Thus equation (7.1) signifies that the sum of the forces on all the particles in a closed system is zero:

$$\sum_a \mathbf{F}_a = 0. \qquad (7.4)$$

In particular, for a system of only two particles, $\mathbf{F}_1 + \mathbf{F}_2 = 0$: the force exerted by the first particle on the second is equal in magnitude, and opposite in direction, to that exerted by the second particle on the first. This is the equality of action and reaction (*Newton's third law*).

If the motion is described by generalised co-ordinates q_i, the derivatives of the Lagrangian with respect to the generalised velocities

$$p_i = \partial L/\partial \dot{q}_i \qquad (7.5)$$

are called *generalised momenta*, and its derivatives with respect to the generalised co-ordinates

$$F_i = \partial L/\partial q_i \qquad (7.6)$$

are called *generalised forces*. In this notation, Lagrange's equations are

$$\dot{p}_i = F_i. \qquad (7.7)$$

In Cartesian co-ordinates the generalised momenta are the components of the vectors $\mathbf{p}_a$. In general, however, the p_i are linear homogeneous functions of the generalised velocities $\dot{q}_i$, and do not reduce to products of mass and velocity.

PROBLEM

A particle of mass m moving with velocity $\mathbf{v}_1$ leaves a half-space in which its potential energy is a constant U_1 and enters another in which its potential energy is a different constant U_2. Determine the change in the direction of motion of the particle.

SOLUTION. The potential energy is independent of the co-ordinates whose axes are parallel to the plane separating the half-spaces. The component of momentum in that plane is therefore conserved. Denoting by θ_1 and θ_2 the angles between the normal to the plane and the velocities $\mathbf{v}_1$ and $\mathbf{v}_2$ of the particle before and after passing the plane, we have $v_1 \sin \theta_1 = v_2 \sin \theta_2$. The relation between v_1 and v_2 is given by the law of conservation of energy, and the result is

$$\frac{\sin \theta_1}{\sin \theta_2} = \sqrt{\left[1 + \frac{2}{mv_1{}^2}(U_1 - U_2)\right]}.$$

§8. Centre of mass

The momentum of a closed mechanical system has different values in different (inertial) frames of reference. If a frame K' moves with velocity $\mathbf{V}$

relative to another frame K, then the velocities $\mathbf{v}_a{}'$ and $\mathbf{v}_a$ of the particles relative to the two frames are such that $\mathbf{v}_a = \mathbf{v}_a{}' + \mathbf{V}$. The momenta $\mathbf{P}$ and $\mathbf{P}'$ in the two frames are therefore related by

$$\mathbf{P} = \sum_a m_a \mathbf{v}_a = \sum_a m_a \mathbf{v}_a{}' + \mathbf{V} \sum_a m_a,$$

or

$$\mathbf{P} = \mathbf{P}' + \mathbf{V} \sum_a m_a. \tag{8.1}$$

In particular, there is always a frame of reference K' in which the total momentum is zero. Putting $\mathbf{P}' = 0$ in (8.1), we find the velocity of this frame:

$$\mathbf{V} = \mathbf{P} / \sum m_a = \sum m_a \mathbf{v}_a / \sum m_a. \tag{8.2}$$

If the total momentum of a mechanical system in a given frame of reference is zero, it is said to be *at rest* relative to that frame. This is a natural generalisation of the term as applied to a particle. Similarly, the velocity $\mathbf{V}$ given by (8.2) is the velocity of the "motion as a whole" of a mechanical system whose momentum is not zero. Thus we see that the law of conservation of momentum makes possible a natural definition of rest and velocity, as applied to a mechanical system as a whole.

Formula (8.2) shows that the relation between the momentum $\mathbf{P}$ and the velocity $\mathbf{V}$ of the system is the same as that between the momentum and velocity of a single particle of mass $\mu = \sum m_a$, the sum of the masses of the particles in the system. This result can be regarded as expressing the *additivity of mass*.

The right-hand side of formula (8.2) can be written as the total time derivative of the expression

$$\mathbf{R} \equiv \sum m_a \mathbf{r}_a / \sum m_a. \tag{8.3}$$

We can say that the velocity of the system as a whole is the rate of motion in space of the point whose radius vector is (8.3). This point is called the *centre of mass* of the system.

The law of conservation of momentum for a closed system can be formulated as stating that the centre of mass of the system moves uniformly in a straight line. In this form it generalises the law of inertia derived in §3 for a single free particle, whose "centre of mass" coincides with the particle itself.

In considering the mechanical properties of a closed system it is natural to use a frame of reference in which the centre of mass is at rest. This eliminates a uniform rectilinear motion of the system as a whole, but such motion is of no interest.

The energy of a mechanical system which is at rest as a whole is usually called its *internal energy* E_i. This includes the kinetic energy of the relative motion of the particles in the system and the potential energy of their interaction. The total energy of a system moving as a whole with velocity V can be written

$$E = \tfrac{1}{2} \mu V^2 + E_i. \tag{8.4}$$

Although this formula is fairly obvious, we may give a direct proof of it. The energies E and E' of a mechanical system in two frames of reference K and K' are related by

$$E = \tfrac{1}{2} \sum_a m_a v_a{}^2 + U$$

$$= \tfrac{1}{2} \sum_a m_a (\mathbf{v}_a' + \mathbf{V})^2 + U$$

$$= \tfrac{1}{2}\mu V^2 + \mathbf{V} \cdot \sum_a m_a \mathbf{v}_a' + \tfrac{1}{2} \sum_a m_a v_a'^2 + U$$

$$= E' + \mathbf{V} \cdot \mathbf{P}' + \tfrac{1}{2}\mu V^2. \tag{8.5}$$

This formula gives the law of transformation of energy from one frame to another, corresponding to formula (8.1) for momentum. If the centre of mass is at rest in K', then $\mathbf{P}' = 0$, $E' = E_i$, and we have (8.4).

PROBLEM

Find the law of transformation of the action S from one inertial frame to another.

SOLUTION. The Lagrangian is equal to the difference of the kinetic and potential energies, and is evidently transformed in accordance with a formula analogous to (8.5):

$$L = L' + \mathbf{V} \cdot \mathbf{P}' + \tfrac{1}{2}\mu V^2.$$

Integrating this with respect to time, we obtain the required law of transformation of the action:

$$S = S' + \mu\mathbf{V} \cdot \mathbf{R}' + \tfrac{1}{2}\mu V^2 t,$$

where $\mathbf{R}'$ is the radius vector of the centre of mass in the frame K'.

§9. Angular momentum

Let us now derive the conservation law which follows from the *isotropy of space*. This isotropy means that the mechanical properties of a closed system do not vary when it is rotated as a whole in any manner in space. Let us therefore consider an infinitesimal rotation of the system, and obtain the condition for the Lagrangian to remain unchanged.

We shall use the vector $\delta\boldsymbol{\phi}$ of the infinitesimal rotation, whose magnitude is the angle of rotation $\delta\phi$, and whose direction is that of the axis of rotation (the direction of rotation being that of a right-handed screw driven along $\delta\boldsymbol{\phi}$).

Let us find, first of all, the resulting increment in the radius vector from an origin on the axis to any particle in the system undergoing rotation. The linear displacement of the end of the radius vector is related to the angle by $|\delta\mathbf{r}| = r \sin\theta \, \delta\phi$ (Fig. 5). The direction of $\delta\mathbf{r}$ is perpendicular to the plane of $\mathbf{r}$ and $\delta\boldsymbol{\phi}$. Hence it is clear that

$$\delta\mathbf{r} = \delta\boldsymbol{\phi} \times \mathbf{r}. \tag{9.1}$$

When the system is rotated, not only the radius vectors but also the velocities of the particles change direction, and all vectors are transformed in the same manner. The velocity increment relative to a fixed system of co-ordinates is

$$\delta\mathbf{v} = \delta\boldsymbol{\phi}\times\mathbf{v}. \tag{9.2}$$

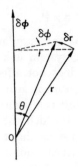

FIG. 5

If these expressions are substituted in the condition that the Lagrangian is unchanged by the rotation:

$$\delta L = \sum_a \left(\frac{\partial L}{\partial\mathbf{r}_a}\cdot\delta\mathbf{r}_a + \frac{\partial L}{\partial\mathbf{v}_a}\cdot\delta\mathbf{v}_a\right) = 0$$

and the derivative $\partial L/\partial\mathbf{v}_a$ replaced by $\mathbf{p}_a$, and $\partial L/\partial\mathbf{r}_a$ by $\dot{\mathbf{p}}_a$, the result is

$$\sum_a(\dot{\mathbf{p}}_a\cdot\delta\boldsymbol{\phi}\times\mathbf{r}_a + \mathbf{p}_a\cdot\delta\boldsymbol{\phi}\times\mathbf{v}_a) = 0$$

or, permuting the factors and taking $\delta\boldsymbol{\phi}$ outside the sum,

$$\delta\boldsymbol{\phi}\sum_a(\mathbf{r}_a\times\dot{\mathbf{p}}_a + \mathbf{v}_a\times\mathbf{p}_a) = \delta\boldsymbol{\phi}\cdot\frac{d}{dt}\sum_a\mathbf{r}_a\times\mathbf{p}_a = 0.$$

Since $\delta\boldsymbol{\phi}$ is arbitrary, it follows that $(d/dt)\sum\mathbf{r}_a\times\mathbf{p}_a = 0$, and we conclude that the vector

$$\mathbf{M} \equiv \sum_a\mathbf{r}_a\times\mathbf{p}_a, \tag{9.3}$$

called the *angular momentum* or *moment of momentum* of the system, is conserved in the motion of a closed system. Like the linear momentum, it is additive, whether or not the particles in the system interact.

There are no other additive integrals of the motion. Thus every closed system has seven such integrals: energy, three components of momentum, and three components of angular momentum.

Since the definition of angular momentum involves the radius vectors of the particles, its value depends in general on the choice of origin. The radius vectors $\mathbf{r}_a$ and $\mathbf{r}_a'$ of a given point relative to origins at a distance $\mathbf{a}$ apart are related by $\mathbf{r}_a = \mathbf{r}_a' + \mathbf{a}$. Hence

$$
\begin{aligned}
\mathbf{M} &= \sum_a \mathbf{r}_a \times \mathbf{p}_a \\
&= \sum_a \mathbf{r}_a' \times \mathbf{p}_a + \mathbf{a} \times \sum_a \mathbf{p}_a \\
&= \mathbf{M}' + \mathbf{a} \times \mathbf{P}.
\end{aligned}
\tag{9.4}
$$

It is seen from this formula that the angular momentum depends on the choice of origin except when the system is at rest as a whole (i.e. $\mathbf{P} = 0$). This indeterminacy, of course, does not affect the law of conservation of angular momentum, since momentum is also conserved in a closed system.

We may also derive a relation between the angular momenta in two inertial frames of reference K and K', of which the latter moves with velocity $\mathbf{V}$ relative to the former. We shall suppose that the origins in the frames K and K' coincide at a given instant. Then the radius vectors of the particles are the same in the two frames, while their velocities are related by $\mathbf{v}_a = \mathbf{v}_a' + \mathbf{V}$. Hence we have

$$
\mathbf{M} = \sum_a m_a \mathbf{r}_a \times \mathbf{v}_a = \sum_a m_a \mathbf{r}_a \times \mathbf{v}_a' + \sum_a m_a \mathbf{r}_a \times \mathbf{V}.
$$

The first sum on the right-hand side is the angular momentum $\mathbf{M}'$ in the frame K'; using in the second sum the radius vector of the centre of mass (8.3), we obtain

$$
\mathbf{M} = \mathbf{M}' + \mu \mathbf{R} \times \mathbf{V}.
\tag{9.5}
$$

This formula gives the law of transformation of angular momentum from one frame to another, corresponding to formula (8.1) for momentum and (8.5) for energy.

If the frame K' is that in which the system considered is at rest as a whole, then $\mathbf{V}$ is the velocity of its centre of mass, $\mu \mathbf{V}$ its total momentum $\mathbf{P}$ relative to K, and

$$
\mathbf{M} = \mathbf{M}' + \mathbf{R} \times \mathbf{P}.
\tag{9.6}
$$

In other words, the angular momentum $\mathbf{M}$ of a mechanical system consists of its "intrinsic angular momentum" in a frame in which it is at rest, and the angular momentum $\mathbf{R} \times \mathbf{P}$ due to its motion as a whole.

Although the law of conservation of all three components of angular momentum (relative to an arbitrary origin) is valid only for a closed system, the law of conservation may hold in a more restricted form even for a system in an external field. It is evident from the above derivation that the component of angular momentum along an axis about which the field is symmetrical is always conserved, for the mechanical properties of the system are unaltered by any rotation about that axis. Here the angular momentum must, of course, be defined relative to an origin lying on the axis.

The most important such case is that of a *centrally symmetric field* or *central field*, i.e. one in which the potential energy depends only on the distance from some particular point (the *centre*). It is evident that the component of angular momentum along any axis passing through the centre is conserved in motion in such a field. In other words, the angular momentum **M** is conserved provided that it is defined with respect to the centre of the field.

Another example is that of a homogeneous field in the z-direction; in such a field, the component M_z of the angular momentum is conserved, whichever point is taken as the origin.

The component of angular momentum along any axis (say the z-axis) can be found by differentiation of the Lagrangian:

$$M_z = \sum_a \frac{\partial L}{\partial \dot{\phi}_a}, \tag{9.7}$$

where the co-ordinate ϕ is the angle of rotation about the z-axis. This is evident from the above proof of the law of conservation of angular momentum, but can also be proved directly. In cylindrical co-ordinates r, ϕ, z we have (substituting $x_a = r_a \cos \phi_a$, $y_a = r_a \sin \phi_a$)

$$M_z = \sum_a m_a(x_a \dot{y}_a - y_a \dot{x}_a)$$

$$= \sum_a m_a r_a^2 \dot{\phi}_a. \tag{9.8}$$

The Lagrangian is, in terms of these co-ordinates,

$$L = \tfrac{1}{2} \sum_a m_a(\dot{r}_a^2 + r_a^2 \dot{\phi}_a^2 + \dot{z}_a^2) - U,$$

and substitution of this in (9.7) gives (9.8).

PROBLEMS

PROBLEM 1. Obtain expressions for the Cartesian components and the magnitude of the angular momentum of a particle in cylindrical co-ordinates r, ϕ, z.

SOLUTION. $M_x = m(r\dot{z} - z\dot{r}) \sin \phi - mrz\dot{\phi} \cos \phi$,
$M_y = -m(r\dot{z} - z\dot{r}) \cos \phi - mrz\dot{\phi} \sin \phi$,
$M_z = mr^2\dot{\phi}$,
$M^2 = m^2 r^2 \dot{\phi}^2 (r^2 + z^2) + m^2(r\dot{z} - z\dot{r})^2$.

PROBLEM 2. The same as Problem 1, but in spherical co-ordinates r, θ, ϕ.

SOLUTION. $M_x = -mr^2(\dot{\theta} \sin \phi + \dot{\phi} \sin \theta \cos \theta \cos \phi)$,
$M_y = mr^2(\dot{\theta} \cos \phi - \dot{\phi} \sin \theta \cos \theta \sin \phi)$,
$M_z = mr^2\dot{\phi} \sin^2\theta$,
$M^2 = m^2 r^4(\dot{\theta}^2 + \dot{\phi}^2 \sin^2\theta)$.

PROBLEM 3. Which components of momentum **P** and angular momentum **M** are conserved in motion in the following fields?

(a) the field of an infinite homogeneous plane, (b) that of an infinite homogeneous cylinder, (c) that of an infinite homogeneous prism, (d) that of two points, (e) that of an infinite homogeneous half-plane, (f) that of a homogeneous cone, (g) that of a homogeneous circular torus, (h) that of an infinite homogeneous cylindrical helix.

SOLUTION. (a) P_x, P_y, M_z (if the plane is the xy-plane), (b) M_z, P_z (if the axis of the cylinder is the z-axis), (c) P_z (if the edges of the prism are parallel to the z-axis), (d) M_z (if the line joining the points is the z-axis), (e) P_y (if the edge of the half-plane is the y-axis), (f) M_z (if the axis of the cone is the z-axis), (g) M_z (if the axis of the torus is the z-axis), (h) the Lagrangian is unchanged by a rotation through an angle $\delta\phi$ about the axis of the helix (let this be the z-axis) together with a translation through a distance $h\delta\phi/2\pi$ along the axis (h being the pitch of the helix). Hence $\delta L = \delta z\, \partial L/\partial z +$ $+\delta\phi\, \partial L/\partial\phi = \delta\phi(h\dot{P}_z/2\pi + \dot{M}_z) = 0$, so that $M_z + hP_z/2\pi = $ constant.

§10. Mechanical similarity

Multiplication of the Lagrangian by any constant clearly does not affect the equations of motion. This fact (already mentioned in §2) makes possible, in a number of important cases, some useful inferences concerning the properties of the motion, without the necessity of actually integrating the equations.

Such cases include those where the potential energy is a homogeneous function of the co-ordinates, i.e. satisfies the condition

$$U(\alpha \mathbf{r}_1, \alpha \mathbf{r}_2, ..., \alpha \mathbf{r}_n) = \alpha^k U(\mathbf{r}_1, \mathbf{r}_2, ..., \mathbf{r}_n), \tag{10.1}$$

where α is any constant and k the degree of homogeneity of the function.

Let us carry out a transformation in which the co-ordinates are changed by a factor α and the time by a factor β: $\mathbf{r}_a \to \alpha \mathbf{r}_a$, $t \to \beta t$. Then all the velocities $\mathbf{v}_a = d\mathbf{r}_a/dt$ are changed by a factor α/β, and the kinetic energy by a factor α^2/β^2. The potential energy is multiplied by α^k. If α and β are such that $\alpha^2/\beta^2 = \alpha^k$, i.e. $\beta = \alpha^{1-\frac{1}{2}k}$, then the result of the transformation is to multiply the Lagrangian by the constant factor α^k, i.e. to leave the equations of motion unaltered.

A change of all the co-ordinates of the particles by the same factor corresponds to the replacement of the paths of the particles by other paths, geometrically similar but differing in size. Thus we conclude that, if the potential energy of the system is a homogeneous function of degree k in the (Cartesian) co-ordinates, the equations of motion permit a series of geometrically similar paths, and the times of the motion between corresponding points are in the ratio

$$t'/t = (l'/l)^{1-\frac{1}{2}k}, \tag{10.2}$$

where l'/l is the ratio of linear dimensions of the two paths. Not only the times but also any mechanical quantities at corresponding points at corresponding times are in a ratio which is a power of l'/l. For example, the velocities, energies and angular momenta are such that

$$v'/v = (l'/l)^{\frac{1}{2}k}, \qquad E'/E = (l'/l)^k, \qquad M'/M = (l'/l)^{1+\frac{1}{2}k}. \tag{10.3}$$

The following are some examples of the foregoing.

As we shall see later, in *small oscillations* the potential energy is a quadratic function of the co-ordinates ($k = 2$). From (10.2) we find that the period of such oscillations is independent of their amplitude.

In a uniform field of force, the potential energy is a linear function of the co-ordinates (see (5.8)), i.e. $k = 1$. From (10.2) we have $t'/t = \sqrt{(l'/l)}$. Hence, for example, it follows that, in fall under gravity, the time of fall is as the square root of the initial altitude.

In the Newtonian attraction of two masses or the Coulomb interaction of two charges, the potential energy is inversely proportional to the distance apart, i.e. it is a homogeneous function of degree $k = -1$. Then $t'/t = (l'/l)^{3/2}$, and we can state, for instance, that the square of the time of revolution in the orbit is as the cube of the size of the orbit (*Kepler's third law*).

If the potential energy is a homogeneous function of the co-ordinates and the motion takes place in a finite region of space, there is a very simple relation between the time average values of the kinetic and potential energies, known as the *virial theorem*.

Since the kinetic energy T is a quadratic function of the velocities, we have by Euler's theorem on homogeneous functions $\sum \mathbf{v}_a \cdot \partial T/\partial \mathbf{v}_a = 2T$, or, putting $\partial T/\partial \mathbf{v}_a = \mathbf{p}_a$, the momentum,

$$2T = \sum_a \mathbf{p}_a \cdot \mathbf{v}_a = \frac{d}{dt}(\sum_a \mathbf{p}_a \cdot \mathbf{r}_a) - \sum_a \mathbf{r}_a \cdot \dot{\mathbf{p}}_a. \tag{10.4}$$

Let us average this equation with respect to time. The average value of any function of time $f(t)$ is defined as

$$\bar{f} = \lim_{\tau \to \infty} \frac{1}{\tau} \int_0^\tau f(t) \, dt.$$

It is easy to see that, if $f(t)$ is the time derivative $dF(t)/dt$ of a bounded function $F(t)$, its mean value is zero. For

$$\bar{f} = \lim_{\tau \to \infty} \frac{1}{\tau} \int_0^\tau \frac{dF}{dt} \, dt = \lim_{\tau \to \infty} \frac{F(\tau) - F(0)}{\tau} = 0.$$

Let us assume that the system executes a motion in a finite region of space and with finite velocities. Then $\sum \mathbf{p}_a \cdot \mathbf{r}_a$ is bounded, and the mean value of the first term on the right-hand side of (10.4) is zero. In the second term we replace $\dot{\mathbf{p}}_a$ by $-\partial U/\partial \mathbf{r}_a$ in accordance with Newton's equations (5.3), obtaining†

$$2\bar{T} = \overline{\sum_a \mathbf{r}_a \cdot \partial U/\partial \mathbf{r}_a}. \tag{10.5}$$

If the potential energy is a homogeneous function of degree k in the radius vectors $\mathbf{r}_a$, then by Euler's theorem equation (10.5) becomes the required relation:

$$2\bar{T} = k\bar{U}. \tag{10.6}$$

---

† The expression on the right of (10.5) is sometimes called the *virial* of the system.

Since $T + \bar{U} = \bar{E} = E$, the relation (10.6) can also be expressed as

$$\bar{U} = 2E/(k+2), \qquad \bar{T} = kE/(k+2), \qquad (10.7)$$

which express $\bar{U}$ and $\bar{T}$ in terms of the total energy of the system.

In particular, for small oscillations ($k = 2$) we have $\bar{T} = \bar{U}$, i.e. the mean values of the kinetic and potential energies are equal. For a Newtonian interaction ($k = -1$) $2\bar{T} = -\bar{U}$, and $E = -\bar{T}$, in accordance with the fact that, in such an interaction, the motion takes place in a finite region of space only if the total energy is negative (see §15).

PROBLEMS

PROBLEM 1. Find the ratio of the times in the same path for particles having different masses but the same potential energy.

SOLUTION. $t'/t = \sqrt{(m'/m)}$.

PROBLEM 2. Find the ratio of the times in the same path for particles having the same mass but potential energies differing by a constant factor.

SOLUTION. $t'/t = \sqrt{(U/U')}$.

CHAPTER III

INTEGRATION OF THE EQUATIONS OF MOTION

§11. Motion in one dimension

THE motion of a system having one degree of freedom is said to take place *in one dimension*. The most general form of the Lagrangian of such a system in fixed external conditions is

$$L = \tfrac{1}{2}a(q)\dot{q}^2 - U(q), \tag{11.1}$$

where $a(q)$ is some function of the generalised co-ordinate q. In particular, if q is a Cartesian co-ordinate (x, say) then

$$L = \tfrac{1}{2}m\dot{x}^2 - U(x). \tag{11.2}$$

The equations of motion corresponding to these Lagrangians can be integrated in a general form. It is not even necessary to write down the equation of motion; we can start from the first integral of this equation, which gives the law of conservation of energy. For the Lagrangian (11.2) (e.g.) we have $\tfrac{1}{2}m\dot{x}^2 + U(x) = E$. This is a first-order differential equation, and can be integrated immediately. Since $dx/dt = \sqrt{\{2[E - U(x)]/m\}}$, it follows that

$$t = \sqrt{(\tfrac{1}{2}m)} \int \frac{dx}{\sqrt{[E - U(x)]}} + \text{constant}. \tag{11.3}$$

The two arbitrary constants in the solution of the equations of motion are here represented by the total energy E and the constant of integration.

Since the kinetic energy is essentially positive, the total energy always exceeds the potential energy, i.e. the motion can take place only in those regions of space where $U(x) < E$. For example, let the function $U(x)$ be of the form shown in Fig. 6 (p. 26). If we draw in the figure a horizontal line corresponding to a given value of the total energy, we immediately find the possible regions of motion. In the example of Fig. 6, the motion can occur only in the range AB or in the range to the right of C.

The points at which the potential energy equals the total energy,

$$U(x) = E, \tag{11.4}$$

give the limits of the motion. They are *turning points*, since the velocity there is zero. If the region of the motion is bounded by two such points, then the motion takes place in a finite region of space, and is said to be *finite*. If the region of the motion is limited on only one side, or on neither, then the motion is *infinite* and the particle goes to infinity.

A finite motion in one dimension is oscillatory, the particle moving repeatedly back and forth between two points (in Fig. 6, in the *potential well AB* between the points x_1 and x_2). The period T of the oscillations, i.e. the time during which the particle passes from x_1 to x_2 and back, is twice the time from x_1 to x_2 (because of the reversibility property, §5) or, by (11.3),

$$T(E) = \sqrt{(2m)} \int_{x_1(E)}^{x_2(E)} \frac{\mathrm{d}x}{\sqrt{[E - U(x)]}}, \tag{11.5}$$

where x_1 and x_2 are roots of equation (11.4) for the given value of E. This formula gives the period of the motion as a function of the total energy of the particle.

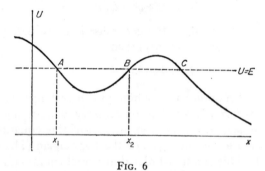

FIG. 6

PROBLEMS

PROBLEM 1. Determine the period of oscillations of a simple pendulum (a particle of mass m suspended by a string of length l in a gravitational field) as a function of the amplitude of the oscillations.

SOLUTION. The energy of the pendulum is $E = \frac{1}{2}ml^2\dot{\phi}^2 - mgl \cos \phi = -mgl \cos \phi_0$, where ϕ is the angle between the string and the vertical, and ϕ_0 the maximum value of ϕ. Calculating the period as the time required to go from $\phi = 0$ to $\phi = \phi_0$, multiplied by four, we find

$$T = 4\sqrt{\frac{l}{2g}} \int_0^{\phi_0} \frac{\mathrm{d}\phi}{\sqrt{(\cos \phi - \cos \phi_0)}}$$

$$= 2\sqrt{\frac{l}{g}} \int_0^{\phi_0} \frac{\mathrm{d}\phi}{\sqrt{(\sin^2\frac{1}{2}\phi_0 - \sin^2\frac{1}{2}\phi)}}.$$

The substitution $\sin \xi = \sin \frac{1}{2}\phi / \sin \frac{1}{2}\phi_0$ converts this to $T = 4\sqrt{(l/g)}K(\sin \frac{1}{2}\phi_0)$, where

$$K(k) = \int_0^{\frac{1}{2}\pi} \frac{\mathrm{d}\xi}{\sqrt{(1 - k^2 \sin^2\xi)}}$$

is the *complete elliptic integral of the first kind.* For $\sin \frac{1}{2}\phi_0 \approx \frac{1}{2}\phi_0 \ll 1$ (small oscillations), an expansion of the function K gives

$$T = 2\pi\sqrt{(l/g)}(1 + \tfrac{1}{16}\phi_0^2 + ...).$$

The first term corresponds to the familiar formula.

PROBLEM 2. Determine the period of oscillation, as a function of the energy, when a particle of mass m moves in fields for which the potential energy is

(a) $U = A|x|^n$,   (b) $U = -U_0/\cosh^2\alpha x$, $-U_0 < E < 0$,   (c) $U = U_0\tan^2\alpha x$.

SOLUTION. (a):

$$T = 2\sqrt{(2m)} \int\limits_0^{(E/A)^{1/n}} \frac{dx}{\sqrt{(E-Ax^n)}}$$

$$= 2\sqrt{\frac{2m}{E}} \cdot \left(\frac{E}{A}\right)^{1/n} \int\limits_0^1 \frac{dy}{\sqrt{(1-y^n)}}.$$

By the substitution $y^n = u$ the integral is reduced to a beta function, which can be expressed in terms of gamma functions:

$$T = \frac{2}{n}\sqrt{\frac{2\pi m}{E}} \cdot \left(\frac{E}{A}\right)^{1/n} \frac{\Gamma(1/n)}{\Gamma(\frac{1}{2}+1/n)}.$$

The dependence of T on E is in accordance with the law of mechanical similarity (10.2), (10.3).

(b) $T = (\pi/\alpha)\sqrt{(2m/|E|)}$.
(c) $T = (\pi/\alpha)\sqrt{[2m/(E+U_0)]}$.

§12. Determination of the potential energy from the period of oscillation

Let us consider to what extent the form of the potential energy $U(x)$ of a field in which a particle is oscillating can be deduced from a knowledge of the period of oscillation T as a function of the energy E. Mathematically, this involves the solution of the integral equation (11.5), in which $U(x)$ is regarded as unknown and $T(E)$ as known.

We shall assume that the required function $U(x)$ has only one minimum in the region of space considered, leaving aside the question whether there exist solutions of the integral equation which do not meet this condition. For convenience, we take the origin at the position of minimum potential energy, and take this minimum energy to be zero (Fig. 7).

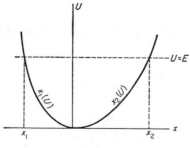

FIG. 7

In the integral (11.5) we regard the co-ordinate x as a function of U. The function $x(U)$ is two-valued: each value of the potential energy corresponds to two different values of x. Accordingly, the integral (11.5) must be divided into two parts before replacing dx by $(dx/dU)\,dU$: one from $x = x_1$ to $x = 0$ and the other from $x = 0$ to $x = x_2$. We shall write the function $x(U)$ in these two ranges as $x = x_1(U)$ and $x = x_2(U)$ respectively.

The limits of integration with respect to U are evidently E and 0, so that we have

$$T(E) = \sqrt{(2m)} \int_0^E \frac{dx_2(U)}{dU} \frac{dU}{\sqrt{(E-U)}} + \sqrt{(2m)} \int_E^0 \frac{dx_1(U)}{dU} \frac{dU}{\sqrt{(E-U)}}$$

$$= \sqrt{(2m)} \int_0^E \left[\frac{dx_2}{dU} - \frac{dx_1}{dU}\right] \frac{dU}{\sqrt{(E-U)}}.$$

If both sides of this equation are divided by $\sqrt{(\alpha-E)}$, where α is a parameter, and integrated with respect to E from 0 to α, the result is

$$\int_0^\alpha \frac{T(E)\,dE}{\sqrt{(\alpha-E)}} = \sqrt{(2m)} \int_0^\alpha \int_0^E \left[\frac{dx_2}{dU} - \frac{dx_1}{dU}\right] \frac{dU\,dE}{\sqrt{[(\alpha-E)(E-U)]}}$$

or, changing the order of integration,

$$\int_0^\alpha \frac{T(E)\,dE}{\sqrt{(\alpha-E)}} = \sqrt{(2m)} \int_0^\alpha \left[\frac{dx_2}{dU} - \frac{dx_1}{dU}\right] dU \int_U^\alpha \frac{dE}{\sqrt{[(\alpha-E)(E-U)]}}.$$

The integral over E is elementary; its value is π. The integral over U is thus trivial, and we have

$$\int_0^\alpha \frac{T(E)\,dE}{\sqrt{(\alpha-E)}} = \pi\sqrt{(2m)}[x_2(\alpha) - x_1(\alpha)],$$

since $x_2(0) = x_1(0) = 0$. Writing U in place of α, we obtain the final result:

$$x_2(U) - x_1(U) = \frac{1}{\pi\sqrt{(2m)}} \int_0^U \frac{T(E)\,dE}{\sqrt{(U-E)}}. \qquad (12.1)$$

Thus the known function $T(E)$ can be used to determine the difference $x_2(U) - x_1(U)$. The functions $x_2(U)$ and $x_1(U)$ themselves remain indeterminate. This means that there is not one but an infinity of curves $U = U(x)$

which give the prescribed dependence of period on energy, and differ in such a way that the difference between the two values of x corresponding to each value of U is the same for every curve.

The indeterminacy of the solution is removed if we impose the condition that the curve $U = U(x)$ must be symmetrical about the U-axis, i.e. that $x_2(U) = -x_1(U) \equiv x(U)$. In this case, formula (12.1) gives for $x(U)$ the unique expression

$$x(U) = \frac{1}{2\pi\sqrt{(2m)}} \int_0^U \frac{T(E)\,dE}{\sqrt{(U-E)}}. \qquad (12.2)$$

§13. The reduced mass

A complete general solution can be obtained for an extremely important problem, that of the motion of a system consisting of two interacting particles (the *two-body problem*).

As a first step towards the solution of this problem, we shall show how it can be considerably simplified by separating the motion of the system into the motion of the centre of mass and that of the particles relative to the centre of mass.

The potential energy of the interaction of two particles depends only on the distance between them, i.e. on the magnitude of the difference in their radius vectors. The Lagrangian of such a system is therefore

$$L = \tfrac{1}{2}m_1\dot{\mathbf{r}}_1^2 + \tfrac{1}{2}m_2\dot{\mathbf{r}}_2^2 - U(|\mathbf{r}_1 - \mathbf{r}_2|). \qquad (13.1)$$

Let $\mathbf{r} \equiv \mathbf{r}_1 - \mathbf{r}_2$ be the relative position vector, and let the origin be at the centre of mass, i.e. $m_1\mathbf{r}_1 + m_2\mathbf{r}_2 = 0$. These two equations give

$$\mathbf{r}_1 = m_2\mathbf{r}/(m_1+m_2), \qquad \mathbf{r}_2 = -m_1\mathbf{r}/(m_1+m_2). \qquad (13.2)$$

Substitution in (13.1) gives

$$L = \tfrac{1}{2}m\dot{\mathbf{r}}^2 - U(r), \qquad (13.3)$$

where

$$m = m_1m_2/(m_1+m_2) \qquad (13.4)$$

is called the *reduced mass*. The function (13.3) is formally identical with the Lagrangian of a particle of mass m moving in an external field $U(r)$ which is symmetrical about a fixed origin.

Thus the problem of the motion of two interacting particles is equivalent to that of the motion of one particle in a given external field $U(r)$. From the solution $\mathbf{r} = \mathbf{r}(t)$ of this problem, the paths $\mathbf{r}_1 = \mathbf{r}_1(t)$ and $\mathbf{r}_2 = \mathbf{r}_2(t)$ of the two particles separately, relative to their common centre of mass, are obtained by means of formulae (13.2).

PROBLEM

A system consists of one particle of mass M and n particles with equal masses m. Eliminate the motion of the centre of mass and so reduce the problem to one involving n particles.

SOLUTION. Let $\mathbf{R}$ be the radius vector of the particle of mass M, and $\mathbf{R}_a$ ($a = 1, 2, ..., n$) those of the particles of mass m. We put $\mathbf{r}_a \equiv \mathbf{R}_a - \mathbf{R}$ and take the origin to be at the centre of mass: $M\mathbf{R} + m\Sigma\mathbf{R}_a = 0$. Hence $\mathbf{R} = -(m/\mu)\Sigma\mathbf{r}_a$, where $\mu \equiv M + nm$; $\mathbf{R}_a = \mathbf{R} + \mathbf{r}_a$. Substitution in the Lagrangian $L = \frac{1}{2}M\dot{\mathbf{R}}^2 + \frac{1}{2}m\Sigma\dot{\mathbf{R}}_a^2 - U$ gives

$$L = \tfrac{1}{2}m\sum_a \mathbf{v}_a^2 - \tfrac{1}{2}(m^2/\mu)\left(\sum_a \mathbf{v}_a\right)^2 - U, \text{ where } \mathbf{v}_a \equiv \dot{\mathbf{r}}_a.$$

The potential energy depends only on the distances between the particles, and so can be written as a function of the $\mathbf{r}_a$.

§14. Motion in a central field

On reducing the two-body problem to one of the motion of a single body, we arrive at the problem of determining the motion of a single particle in an external field such that its potential energy depends only on the distance r from some fixed point. This is called a *central* field. The force acting on the particle is $\mathbf{F} = -\partial U(r)/\partial \mathbf{r} = -(\mathrm{d}U/\mathrm{d}r)\mathbf{r}/r$; its magnitude is likewise a function of r only, and its direction is everywhere that of the radius vector.

As has already been shown in §9, the angular momentum of any system relative to the centre of such a field is conserved. The angular momentum of a single particle is $\mathbf{M} = \mathbf{r} \times \mathbf{p}$. Since $\mathbf{M}$ is perpendicular to $\mathbf{r}$, the constancy of $\mathbf{M}$ shows that, throughout the motion, the radius vector of the particle lies in the plane perpendicular to $\mathbf{M}$.

Thus the path of a particle in a central field lies in one plane. Using polar co-ordinates r, ϕ in that plane, we can write the Lagrangian as

$$L = \tfrac{1}{2}m(\dot{r}^2 + r^2\dot{\phi}^2) - U(r); \tag{14.1}$$

see (4.5). This function does not involve the co-ordinate ϕ explicitly. Any generalised co-ordinate q_i which does not appear explicitly in the Lagrangian is said to be *cyclic*. For such a co-ordinate we have, by Lagrange's equation, $(\mathrm{d}/\mathrm{d}t)\,\partial L/\partial \dot{q}_i = \partial L/\partial q_i = 0$, so that the corresponding generalised momentum $p_i = \partial L/\partial \dot{q}_i$ is an integral of the motion. This leads to a considerable simplification of the problem of integrating the equations of motion when there are cyclic co-ordinates.

In the present case, the generalised momentum $p_\phi = mr^2\dot{\phi}$ is the same as the angular momentum $M_z = M$ (see (9.6)), and we return to the known law of conservation of angular momentum:

$$M = mr^2\dot{\phi} = \text{constant}. \tag{14.2}$$

This law has a simple geometrical interpretation in the plane motion of a single particle in a central field. The expression $\frac{1}{2}r \cdot r\mathrm{d}\phi$ is the area of the sector bounded by two neighbouring radius vectors and an element of the path

(Fig. 8). Calling this area df, we can write the angular momentum of the particle as

$$M = 2m\dot{f}, \tag{14.3}$$

where the derivative $\dot{f}$ is called the *sectorial velocity*. Hence the conservation of angular momentum implies the constancy of the sectorial velocity: in equal times the radius vector of the particle sweeps out equal areas (*Kepler's second law*).†

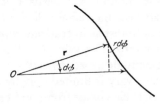

FIG. 8

The complete solution of the problem of the motion of a particle in a central field is most simply obtained by starting from the laws of conservation of energy and angular momentum, without writing out the equations of motion themselves. Expressing $\dot{\phi}$ in terms of M from (14.2) and substituting in the expression for the energy, we obtain

$$E = \tfrac{1}{2}m(\dot{r}^2 + r^2\dot{\phi}^2) + U(r) = \tfrac{1}{2}m\dot{r}^2 + \tfrac{1}{2}M^2/mr^2 + U(r). \tag{14.4}$$

Hence

$$\dot{r} \equiv \frac{dr}{dt} = \sqrt{\left\{\frac{2}{m}[E - U(r)] - \frac{M^2}{m^2r^2}\right\}} \tag{14.5}$$

or, integrating,

$$t = \int dr \Big/ \sqrt{\left\{\frac{2}{m}[E - U(r)] - \frac{M^2}{m^2r^2}\right\}} + \text{constant.} \tag{14.6}$$

Writing (14.2) as d$\phi = M\,dt/mr^2$, substituting dt from (14.5) and integrating, we find

$$\phi = \int \frac{M\,dr/r^2}{\sqrt{\{2m[E - U(r)] - M^2/r^2\}}} + \text{constant.} \tag{14.7}$$

Formulae (14.6) and (14.7) give the general solution of the problem. The latter formula gives the relation between r and ϕ, i.e. the equation of the path. Formula (14.6) gives the distance r from the centre as an implicit function of time. The angle ϕ, it should be noted, always varies monotonically with time, since (14.2) shows that $\dot{\phi}$ can never change sign.

† The law of conservation of angular momentum for a particle moving in a central field is sometimes called the *area integral*.

The expression (14.4) shows that the radial part of the motion can be re-garded as taking place in one dimension in a field where the "effective poten-tial energy" is

$$U_{\text{eff}} = U(r) + M^2/2mr^2. \tag{14.8}$$

The quantity $M^2/2mr^2$ is called the *centrifugal energy*. The values of r for which

$$U(r) + M^2/2mr^2 = E \tag{14.9}$$

determine the limits of the motion as regards distance from the centre. When equation (14.9) is satisfied, the radial velocity $\dot{r}$ is zero. This does not mean that the particle comes to rest as in true one-dimensional motion, since the angular velocity $\dot{\phi}$ is not zero. The value $\dot{r} = 0$ indicates a *turning point* of the path, where $r(t)$ begins to decrease instead of increasing, or *vice versa*.

If the range in which r may vary is limited only by the condition $r \geqslant r_{\min}$, the motion is infinite: the particle comes from, and returns to, infinity.

If the range of r has two limits $r_{\min}$ and $r_{\max}$, the motion is finite and the path lies entirely within the annulus bounded by the circles $r = r_{\max}$ and $r = r_{\min}$. This does not mean, however, that the path must be a closed curve. During the time in which r varies from $r_{\max}$ to $r_{\min}$ and back, the radius vector turns through an angle $\Delta\phi$ which, according to (14.7), is given by

$$\Delta\phi = 2 \int_{r_{\min}}^{r_{\max}} \frac{M \, dr/r^2}{\sqrt{[2m(E - U) - M^2/r^2]}}. \tag{14.10}$$

The condition for the path to be closed is that this angle should be a rational fraction of 2π, i.e. that $\Delta\phi = 2\pi m/n$, where m and n are integers. In that case, after n periods, the radius vector of the particle will have made m complete revolutions and will occupy its original position, so that the path is closed.

Such cases are exceptional, however, and when the form of $U(r)$ is arbitrary the angle $\Delta\phi$ is not a rational fraction of 2π. In general, therefore, the path of a particle executing a finite motion is not closed. It passes through the minimum and maximum distances an infinity of times, and after infinite time it covers the entire annulus between the two bounding circles. The path shown in Fig. 9 is an example.

There are only two types of central field in which all finite motions take place in closed paths. They are those in which the potential energy of the particle varies as $1/r$ or as r^2. The former case is discussed in §15; the latter is that of the *space oscillator* (see §23, Problem 3).

At a turning point the square root in (14.5), and therefore the integrands in (14.6) and (14.7), change sign. If the angle ϕ is measured from the direc-tion of the radius vector to the turning point, the parts of the path on each side of that point differ only in the sign of ϕ for each value of r, i.e. the path is symmetrical about the line $\phi = 0$. Starting, say, from a point where $r = r_{\max}$ the particle traverses a segment of the path as far as a point with $r = r_{\min}$, then follows a symmetrically placed segment to the next point where $r = r_{max}$, and so on. Thus the entire path is obtained by repeating identical segments forwards and backwards. This applies also to infinite paths, which consist of two symmetrical branches extending from the turning point ($r = r_{min}$) to infinity.

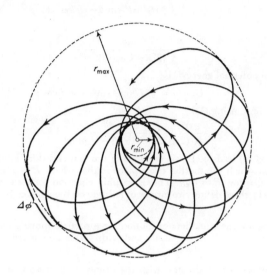

FIG. 9

The presence of the centrifugal energy when $M \neq 0$, which becomes infinite as $1/r^2$ when $r \to 0$, generally renders it impossible for the particle to reach the centre of the field, even if the field is an attractive one. A "fall" of the particle to the centre is possible only if the potential energy tends sufficiently rapidly to $-\infty$ as $r \to 0$. From the inequality

$$\tfrac{1}{2}m\dot{r}^2 = E - U(r) - M^2/2mr^2 > 0,$$

or $r^2 U(r) + M^2/2m < Er^2$, it follows that r can take values tending to zero only if

$$[r^2 U(r)]_{r \to 0} < -M^2/2m, \tag{14.11}$$

i.e. $U(r)$ must tend to $-\infty$ either as $-\alpha/r^2$ with $\alpha > M^2/2m$, or proportionally to $-1/r^n$ with $n > 2$.

PROBLEMS

PROBLEM 1. Integrate the equations of motion for a *spherical pendulum* (a particle of mass m moving on the surface of a sphere of radius l in a gravitational field).

SOLUTION. In spherical co-ordinates, with the origin at the centre of the sphere and the polar axis vertically downwards, the Lagrangian of the pendulum is

$$\tfrac{1}{2}ml^2(\dot{\theta}^2 + \dot{\phi}^2 \sin^2\theta) + mgl \cos \theta.$$

The co-ordinate ϕ is cyclic, and hence the generalised momentum p_ϕ, which is the same as the z-component of angular momentum, is conserved:

$$ml^2\dot\phi\,\sin^2\theta = M_z = \text{constant}. \tag{1}$$

The energy is

$$\begin{aligned}
E &= \tfrac12 ml^2(\dot\theta^2 + \dot\phi^2\sin^2\theta) - mgl\,\cos\theta \\
&= \tfrac12 ml^2\dot\theta^2 + \tfrac12 M_z{}^2/ml^2\sin^2\theta - mgl\,\cos\theta.
\end{aligned} \tag{2}$$

Hence

$$t = \int \frac{d\theta}{\sqrt{\{2[E - U_{\text{eff}}(\theta)]/ml^2\}}}, \tag{3}$$

where the "effective potential energy" is

$$U_{\text{eff}}(\theta) = \tfrac12 M_z{}^2/ml^2\sin^2\theta - mgl\,\cos\theta.$$

For the angle ϕ we find, using (1),

$$\phi = \frac{M_z}{l\sqrt{(2m)}} \int \frac{d\theta}{\sin^2\theta\sqrt{[E - U_{\text{eff}}(\theta)]}}. \tag{4}$$

The integrals (3) and (4) lead to elliptic integrals of the first and third kinds respectively.

The range of θ in which the motion takes place is that where $E > U_{\text{eff}}$, and its limits are given by the equation $E = U_{\text{eff}}$. This is a cubic equation for $\cos\theta$, having two roots between -1 and $+1$; these define two circles of latitude on the sphere, between which the path lies.

PROBLEM 2. Integrate the equations of motion for a particle moving on the surface of a cone (of vertical angle 2α) placed vertically and with vertex downwards in a gravitational field.

SOLUTION. In spherical co-ordinates, with the origin at the vertex of the cone and the polar axis vertically upwards, the Lagrangian is $\tfrac12 m(\dot r^2 + r^2\dot\phi^2\sin^2\alpha) - mgr\,\cos\alpha$. The co-ordinate ϕ is cyclic, and $M_z = mr^2\dot\phi\sin^2\alpha$ is again conserved. The energy is

$$E = \tfrac12 m\dot r^2 + \tfrac12 M_z{}^2/mr^2\sin^2\alpha + mgr\,\cos\alpha.$$

By the same method as in Problem 1, we find

$$t = \int \frac{dr}{\sqrt{\{2[E - U_{\text{eff}}(r)]/m\}}},$$

$$\phi = \frac{M_z}{\sqrt{(2m)}\,\sin^2\alpha} \int \frac{dr}{r^2\sqrt{[E - U_{\text{eff}}(r)]}},$$

$$U_{\text{eff}}(r) = \frac{M_z{}^2}{2mr^2\sin^2\alpha} + mgr\,\cos\alpha.$$

The condition $E = U_{\text{eff}}(r)$ is (if $M_z \neq 0$) a cubic equation for r, having two positive roots; these define two horizontal circles on the cone, between which the path lies.

PROBLEM 3. Integrate the equations of motion for a pendulum of mass m_2, with a mass m_1 at the point of support which can move on a horizontal line lying in the plane in which m_2 moves (Fig. 2, §5).

SOLUTION. In the Lagrangian derived in §5, Problem 2, the co-ordinate x is cyclic. The generalised momentum P_x, which is the horizontal component of the total momentum of the system, is therefore conserved:

$$P_x = (m_1 + m_2)\dot x + m_2 l\dot\phi\,\cos\phi = \text{constant}. \tag{1}$$

The system may always be taken to be at rest as a whole. Then the constant in (1) is zero and integration gives

$$(m_1 + m_2)x + m_2 l\,\sin\phi = \text{constant}, \tag{2}$$

which expresses the fact that the centre of mass of the system does not move horizontally.

Using (1), we find the energy in the form

$$E = \tfrac{1}{2}m_2l^2\dot\phi^2\left(1 - \frac{m_2}{m_1+m_2}\cos^2\phi\right) - m_2gl\cos\phi.\qquad(3)$$

Hence

$$t = l\sqrt{\frac{m_2}{2(m_1+m_2)}}\int\sqrt{\frac{m_1+m_2\sin^2\phi}{E+m_2gl\cos\phi}}\,d\phi.$$

Expressing the co-ordinates $x_2 = x + l\sin\phi$, $y = l\cos\phi$ of the particle m_2 in terms of ϕ by means of (2), we find that its path is an arc of an ellipse with horizontal semi-axis $lm_1/(m_1+m_2)$ and vertical semi-axis l. As $m_1 \to \infty$ we return to the familiar simple pendulum, which moves in an arc of a circle.

§15. Kepler's problem

An important class of central fields is formed by those in which the potential energy is inversely proportional to r, and the force accordingly inversely proportional to r^2. They include the fields of Newtonian gravitational attraction and of Coulomb electrostatic interaction; the latter may be either attractive or repulsive.

Let us first consider an attractive field, where

$$U = -\alpha/r\qquad(15.1)$$

with α a positive constant. The "effective" potential energy

$$U_{\text{eff}} = -\frac{\alpha}{r} + \frac{M^2}{2mr^2}\qquad(15.2)$$

is of the form shown in Fig. 10. As $r \to 0$, U_{eff} tends to $+\infty$, and as $r \to \infty$ it tends to zero from negative values; for $r = M^2/m\alpha$ it has a minimum value

$$U_{\text{eff, min}} = -m\alpha^2/2M^2.\qquad(15.3)$$

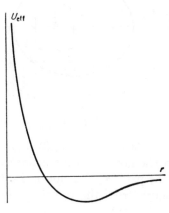

Fig. 10

It is seen at once from Fig. 10 that the motion is finite for $E < 0$ and infinite for $E > 0$.

The shape of the path is obtained from the general formula (14.7). Substituting there $U = -\alpha/r$ and effecting the elementary integration, we have

$$\phi = \cos^{-1}\frac{(M/r)-(m\alpha/M)}{\sqrt{\left(2mE+\dfrac{m^2\alpha^2}{M^2}\right)}}+\text{constant}.$$

Taking the origin of ϕ such that the constant is zero, and putting

$$p = M^2/m\alpha, \qquad e = \sqrt{[1+(2EM^2/m\alpha^2)]}, \tag{15.4}$$

we can write the equation of the path as

$$p/r = 1+e\cos\phi. \tag{15.5}$$

This is the equation of a conic section with one focus at the origin; $2p$ is called the *latus rectum* of the orbit and e the *eccentricity*. Our choice of the origin of ϕ is seen from (15.5) to be such that the point where $\phi = 0$ is the point nearest to the origin (called the *perihelion*).

In the equivalent problem of two particles interacting according to the law (15.1), the orbit of each particle is a conic section, with one focus at the centre of mass of the two particles.

It is seen from (15.4) that, if $E < 0$, then the eccentricity $e < 1$, i.e. the orbit is an ellipse (Fig. 11) and the motion is finite, in accordance with what has been said earlier in this section. According to the formulae of analytical geometry, the major and minor semi-axes of the ellipse are

$$a = p/(1-e^2) = \alpha/2|E|, \qquad b = p/\sqrt{(1-e^2)} = M/\sqrt{(2m|E|)}. \tag{15.6}$$

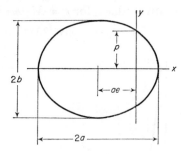

Fig. 11

The least possible value of the energy is (15.3), and then $e = 0$, i.e. the ellipse becomes a circle. It may be noted that the major axis of the ellipse depends only on the energy of the particle, and not on its angular momentum. The least and greatest distances from the centre of the field (the focus of the ellipse) are

$$r_{\min} = p/(1+e) = a(1-e), \qquad r_{\max} = p/(1-e) = a(1+e). \tag{15.7}$$

These expressions, with a and e given by (15.6) and (15.4), can, of course, also be obtained directly as the roots of the equation $U_{\text{eff}}(r) = E$.

The period T of revolution in an elliptical orbit is conveniently found by using the law of conservation of angular momentum in the form of the area integral (14.3). Integrating this equation with respect to time from zero to T, we have $2mf = TM$, where f is the area of the orbit. For an ellipse $f = \pi ab$, and by using the formulae (15.6) we find

$$T = 2\pi a^{3/2}\sqrt{(m/\alpha)}$$

$$= \pi\alpha\sqrt{(m/2|E|^3)}. \tag{15.8}$$

The proportionality between the square of the period and the cube of the linear dimension of the orbit has already been demonstrated in §10. It may also be noted that the period depends only on the energy of the particle.

For $E \geqslant 0$ the motion is infinite. If $E > 0$, the eccentricity $e > 1$, i.e. the the path is a hyperbola with the origin as internal focus (Fig. 12). The distance of the perihelion from the focus is

$$r_{min} = p/(e+1) = a(e-1), \tag{15.9}$$

where $a = p/(e^2-1) = \alpha/2E$ is the "semi-axis" of the hyperbola.

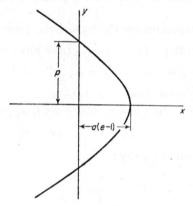

FIG. 12

If $E = 0$, the eccentricity $e = 1$, and the particle moves in a parabola with perihelion distance $r_{min} = \tfrac{1}{2}p$. This case occurs if the particle starts from rest at infinity.

The co-ordinates of the particle as functions of time in the orbit may be found by means of the general formula (14.6). They may be represented in a convenient parametric form as follows.

Let us first consider elliptical orbits. With a and e given by (15.6) and (15.4) we can write the integral (14.6) for the time as

$$t = \sqrt{\frac{m}{2|E|}} \int \frac{r\,dr}{\sqrt{[-r^2+(\alpha/|E|)r-(M^2/2m|E|)]}}$$

$$= \sqrt{\frac{ma}{\alpha}} \int \frac{r\,dr}{\sqrt{[a^2e^2-(r-a)^2]}}.$$

The obvious substitution $r - a = -ae \cos \xi$ converts the integral to

$$t = \sqrt{\frac{ma^3}{\alpha}} \int (1 - e \cos \xi) \, d\xi = \sqrt{\frac{ma^3}{\alpha}} (\xi - e \sin \xi) + \text{constant}.$$

If time is measured in such a way that the constant is zero, we have the following parametric dependence of r on t:

$$r = a(1 - e \cos \xi), \qquad t = \sqrt{(ma^3/\alpha)}(\xi - e \sin \xi), \qquad (15.10)$$

the particle being at perihelion at $t = 0$. The Cartesian co-ordinates $x = r \cos \phi$, $y = r \sin \phi$ (the x and y axes being respectively parallel to the major and minor axes of the ellipse) can likewise be expressed in terms of the parameter ξ. From (15.5) and (15.10) we have

$$ex = p - r = a(1 - e^2) - a(1 - e \cos \xi) = ae(\cos \xi - e);$$

y is equal to $\sqrt{(r^2 - x^2)}$. Thus

$$x = a(\cos \xi - e), \qquad y = a\sqrt{(1 - e^2)} \sin \xi. \qquad (15.11)$$

A complete passage round the ellipse corresponds to an increase of ξ from 0 to 2π.

Entirely similar calculations for the hyperbolic orbits give

$$r = a(e \cosh \xi - 1), \qquad t = \sqrt{(ma^3/\alpha)}(e \sinh \xi - \xi),$$
$$x = a(e - \cosh \xi), \qquad y = a\sqrt{(e^2 - 1)} \sinh \xi, \qquad (15.12)$$

where the parameter ξ varies from $-\infty$ to $+\infty$.

Let us now consider motion in a repulsive field, where

$$U = \alpha/r \qquad (\alpha > 0). \qquad (15.13)$$

Here the effective potential energy is

$$U_{\text{eff}} = \frac{\alpha}{r} + \frac{M^2}{2mr^2}$$

and decreases monotonically from $+\infty$ to zero as r varies from zero to infinity. The energy of the particle must be positive, and the motion is always infinite. The calculations are exactly similar to those for the attractive field. The path is a hyperbola:

$$p/r = -1 + e \cos \phi, \qquad (15.14)$$

where p and e are again given by (15.4). The path passes the centre of the field in the manner shown in Fig. 13. The perihelion distance is

$$r_{\text{min}} = p/(e - 1) = a(e + 1). \qquad (15.15)$$

The time dependence is given by the parametric equations

$$r = a(e \cosh \xi + 1), \qquad t = \sqrt{(ma^3/\alpha)}(e \sinh \xi + \xi),$$
$$x = a(\cosh \xi + e), \qquad y = a\sqrt{(e^2 - 1)} \sinh \xi. \qquad (15.16)$$

To conclude this section, we shall show that there is an integral of the motion which exists only in fields $U = \alpha/r$ (with either sign of α). It is easy to verify by direct calculation that the quantity

$$\mathbf{v} \times \mathbf{M} + \alpha \mathbf{r}/r \qquad (15.17)$$

is constant. For its total time derivative is $\dot{\mathbf{v}} \times \mathbf{M} + \alpha \mathbf{v}/r - \alpha \mathbf{r}(\mathbf{v} \cdot \mathbf{r})/r^3$ or, since $\mathbf{M} = m\mathbf{r} \times \mathbf{v}$,

$$m\mathbf{r}(\mathbf{v} \cdot \dot{\mathbf{v}}) - m\mathbf{v}(\mathbf{r} \cdot \dot{\mathbf{v}}) + \alpha \mathbf{v}/r - \alpha \mathbf{r}(\mathbf{v} \cdot \mathbf{r})/r^3.$$

Putting $m\dot{\mathbf{v}} = \alpha\mathbf{r}/r^3$ from the equation of motion, we find that this expression vanishes.

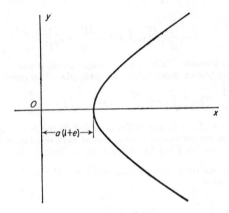

$\leftarrow a(1+e)\rightarrow$

FIG. 13

The direction of the conserved vector (15.17) is along the major axis from the focus to the perihelion, and its magnitude is αe. This is most simply seen by considering its value at perihelion.

It should be emphasised that the integral (15.17) of the motion, like M and E, is a one-valued function of the state (position and velocity) of the particle. We shall see in §50 that the existence of such a further one-valued integral is due to the *degeneracy* of the motion.

PROBLEMS

PROBLEM 1. Find the time dependence of the co-ordinates of a particle with energy $E = 0$ moving in a parabola in a field $U = -\alpha/r$.

SOLUTION. In the integral

$$t = \int \frac{r \, dr}{\sqrt{[(2\alpha/m)r - (M^2/m^2)]}}$$

we substitute $r = M^2(1+\eta^2)/2m\alpha = \tfrac{1}{2}p(1+\eta^2)$, obtaining the following parametric form of the required dependence:

$$r = \tfrac{1}{2}p(1+\eta^2), \qquad t = \sqrt{(mp^3/\alpha)} \cdot \tfrac{1}{2}\eta(1 + \tfrac{1}{3}\eta^2),$$
$$x = \tfrac{1}{2}p(1-\eta^2), \qquad y = p\eta.$$

The parameter η varies from $-\infty$ to $+\infty$.

PROBLEM 2. Integrate the equations of motion for a particle in a central field

$$U = -\alpha/r^2 \qquad (\alpha > 0).$$

SOLUTION. From formulae (14.6) and (14.7) we have, if ϕ and t are appropriately measured,

(a) for $E > 0$ and $M^2/2m > \alpha$, $\dfrac{1}{r} = \sqrt{\dfrac{2mE}{M^2 - 2m\alpha}} \cos\left[\phi\sqrt{\left(1 - \dfrac{2m\alpha}{M^2}\right)}\right]$,

(b) for $E > 0$ and $M^2/2m < \alpha$, $\dfrac{1}{r} = \sqrt{\dfrac{2mE}{2m\alpha - M^2}} \sinh\left[\phi\sqrt{\left(\dfrac{2m\alpha}{M^2} - 1\right)}\right]$,

(c) for $E < 0$ and $M^2/2m < \alpha$, $\dfrac{1}{r} = \sqrt{\dfrac{2m|E|}{2m\alpha - M^2}} \cosh\left[\phi\sqrt{\left(\dfrac{2m\alpha}{M^2} - 1\right)}\right]$.

In all three cases

$$t = \frac{1}{E}\sqrt{\left[\tfrac{1}{2}m\left(Er^2 - \frac{M^2}{2m} + \alpha\right)\right]}.$$

In cases (b) and (c) the particle "falls" to the centre along a path which approaches the origin as $\phi \to \infty$. The fall from a given value of r takes place in a finite time, namely

$$\frac{1}{E}\sqrt{(\tfrac{1}{2}m)}\left\{\sqrt{\left(\alpha - \frac{M^2}{2m} + Er^2\right)} - \sqrt{\left(\alpha - \frac{M^2}{2m}\right)}\right\}.$$

PROBLEM 3. When a small correction $\delta U(r)$ is added to the potential energy $U = -\alpha/r$, the paths of finite motion are no longer closed, and at each revolution the perihelion is displaced through a small angle $\delta\phi$. Find $\delta\phi$ when (a) $\delta U = \beta/r^2$, (b) $\delta U = \gamma/r^3$.

SOLUTION. When r varies from $r_{\min}$ to $r_{\max}$ and back, the angle ϕ varies by an amount (14.10), which we write as

$$\Delta\phi = -2\frac{\partial}{\partial M} \int_{r_{\min}}^{r_{\max}} \sqrt{\left[2m(E - U) - \frac{M^2}{r^2}\right]}\, dr,$$

in order to avoid the occurrence of spurious divergences. We put $U = -\alpha/r + \delta U$, and expand the integrand in powers of δU; the zero-order term in the expansion gives 2π, and the first-order term gives the required change $\delta\phi$:

$$\delta\phi = \frac{\partial}{\partial M} \int_{r_{\min}}^{r_{\max}} \frac{2m\delta U\, dr}{\sqrt{\left[2m\left(E + \dfrac{\alpha}{r}\right) - \dfrac{M^2}{r^2}\right]}} = \frac{\partial}{\partial M}\left(\frac{2m}{M} \int_0^{\pi} r^2 \delta U\, d\phi\right), \tag{1}$$

where we have changed from the integration over r to one over ϕ, along the path of the "unperturbed" motion.

In case (a), the integration in (1) is trivial: $\delta\phi = -2\pi\beta m/M^2 = -2\pi\beta/\alpha p$, where $2p$ (15.4) is the latus rectum of the unperturbed ellipse. In case (b) $r^2\delta U = \gamma/r$ and, with $1/r$ given by (15.5), we have $\delta\phi = -6\pi\alpha\gamma m^2/M^4 = -6\pi\gamma/\alpha p^2$.

CHAPTER IV

COLLISIONS BETWEEN PARTICLES

§16. Disintegration of particles

IN many cases the laws of conservation of momentum and energy alone can be used to obtain important results concerning the properties of various mechanical processes. It should be noted that these properties are independent of the particular type of interaction between the particles involved.

Let us consider a "spontaneous" disintegration (that is, one not due to external forces) of a particle into two "constituent parts", i.e. into two other particles which move independently after the disintegration.

This process is most simply described in a frame of reference in which the particle is at rest before the disintegration. The law of conservation of momentum shows that the sum of the momenta of the two particles formed in the disintegration is then zero; that is, the particles move apart with equal and opposite momenta. The magnitude p_0 of either momentum is given by the law of conservation of energy:

$$E_i = E_{1i} + \frac{p_0^2}{2m_1} + E_{2i} + \frac{p_0^2}{2m_2};$$

here m_1 and m_2 are the masses of the particles, E_{1i} and E_{2i} their internal energies, and E_i the internal energy of the original particle. If ϵ is the "disintegration energy", i.e. the difference

$$\epsilon = E_i - E_{1i} - E_{2i}, \tag{16.1}$$

which must obviously be positive, then

$$\epsilon = \tfrac{1}{2} p_0^2 \left(\frac{1}{m_1} + \frac{1}{m_2} \right) = \frac{p_0^2}{2m}, \tag{16.2}$$

which determines p_0; here m is the reduced mass of the two particles. The velocities are $v_{10} = p_0/m_1$, $v_{20} = p_0/m_2$.

Let us now change to a frame of reference in which the primary particle moves with velocity V before the break-up. This frame is usually called the *laboratory system*, or L system, in contradistinction to the *centre-of-mass system*, or C system, in which the total momentum is zero. Let us consider one of the resulting particles, and let $\mathbf{v}$ and $\mathbf{v}_0$ be its velocities in the L and the C system respectively. Evidently $\mathbf{v} = \mathbf{V} + \mathbf{v}_0$, or $\mathbf{v} - \mathbf{V} = \mathbf{v}_0$, and so

$$v^2 + V^2 - 2vV \cos \theta = v_0^2, \tag{16.3}$$

where θ is the angle at which this particle moves relative to the direction of the velocity $\mathbf{V}$. This equation gives the velocity of the particle as a function of its direction of motion in the L system. In Fig. 14 the velocity $\mathbf{v}$ is represented by a vector drawn to any point on a circle† of radius v_0 from a point A at a distance V from the centre. The cases $V < v_0$ and $V > v_0$ are shown in Figs. 14a, b respectively. In the former case θ can have any value, but in the latter case the particle can move only forwards, at an angle θ which does not exceed $\theta_{\max}$, given by

$$\sin \theta_{\max} = v_0/V; \tag{16.4}$$

this is the direction of the tangent from the point A to the circle.

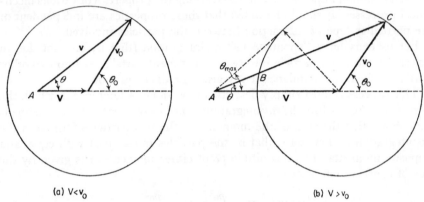

(a) $V < v_0$ (b) $V > v_0$

Fig. 14

The relation between the angles θ and θ_0 in the L and C systems is evidently (Fig. 14)

$$\tan \theta = v_0 \sin \theta_0/(v_0 \cos \theta_0 + V). \tag{16.5}$$

If this equation is solved for $\cos \theta_0$, we obtain

$$\cos \theta_0 = -\frac{V}{v_0} \sin^2\theta \pm \cos \theta \sqrt{\left(1 - \frac{V^2}{v_0^2}\sin^2\theta\right)}. \tag{16.6}$$

For $v_0 > V$ the relation between θ_0 and θ is one-to-one (Fig. 14a). The plus sign must be taken in (16.6), so that $\theta_0 = 0$ when $\theta = 0$. If $v_0 < V$, however, the relation is not one-to-one: for each value of θ there are two values of θ_0, which correspond to vectors $\mathbf{v}_0$ drawn from the centre of the circle to the points B and C (Fig. 14b), and are given by the two signs in (16.6).

In physical applications we are usually concerned with the disintegration of not one but many similar particles, and this raises the problem of the distribution of the resulting particles in direction, energy, etc. We shall assume that the primary particles are randomly oriented in space, i.e. isotropically on average.

---

† More precisely, to any point on a sphere of radius v_0, of which Fig. 14 shows a diametral section.

In the C system, this problem is very easily solved: every resulting particle (of a given kind) has the same energy, and their directions of motion are isotropically distributed. The latter fact depends on the assumption that the primary particles are randomly oriented, and can be expressed by saying that the fraction of particles entering a solid angle element do_0 is proportional to do_0, i.e. equal to $do_0/4\pi$. The distribution with respect to the angle θ_0 is obtained by putting $do_0 = 2\pi \sin \theta_0 \, d\theta_0$, i.e. the corresponding fraction is

$$\tfrac{1}{2} \sin \theta_0 \, d\theta_0. \tag{16.7}$$

The corresponding distributions in the L system are obtained by an appropriate transformation. For example, let us calculate the kinetic energy distribution in the L system. Squaring the equation $\mathbf{v} = \mathbf{v}_0 + \mathbf{V}$, we have $v^2 = v_0^2 + V^2 + 2v_0 V \cos \theta_0$, whence $d(\cos \theta_0) = d(v^2)/2v_0 V$. Using the kinetic energy $T = \tfrac{1}{2}mv^2$, where m is m_1 or m_2 depending on which kind of particle is under consideration, and substituting in (16.7), we find the required distribution:

$$(1/2mv_0 V) \, dT. \tag{16.8}$$

The kinetic energy can take values between $T_{\min} = \tfrac{1}{2}m(v_0 - V)^2$ and $T_{\max} = \tfrac{1}{2}m(v_0 + V)^2$. The particles are, according to (16.8), distributed uniformly over this range.

When a particle disintegrates into more than two parts, the laws of conservation of energy and momentum naturally allow considerably more freedom as regards the velocities and directions of motion of the resulting particles. In particular, the energies of these particles in the C system do not have determinate values. There is, however, an upper limit to the kinetic energy of any one of the resulting particles. To determine the limit, we consider the system formed by all these particles except the one concerned (whose mass is m_1, say), and denote the "internal energy" of that system by E_i'. Then the kinetic energy of the particle m_1 is, by (16.1) and (16.2), $T_{10} = p_0^2/2m_1 = (M - m_1)(E_i - E_{1i} - E_i')/M$, where M is the mass of the primary particle. It is evident that T_{10} has its greatest possible value when E_i' is least. For this to be so, all the resulting particles except m_1 must be moving with the same velocity. Then E_i' is simply the sum of their internal energies, and the difference $E_i - E_{1i} - E_i'$ is the disintegration energy ϵ. Thus

$$T_{10,\max} = (M - m_1)\epsilon/M. \tag{16.9}$$

PROBLEMS

PROBLEM 1. Find the relation between the angles θ_1, θ_2 (in the L system) after a disintegration into two particles.

SOLUTION. In the C system, the corresponding angles are related by $\theta_{10} = \pi - \theta_{20}$. Calling θ_{10} simply θ_0 and using formula (16.5) for each of the two particles, we can put $V + v_{10} \cos \theta_0 = v_{10} \sin \theta_0 \cot \theta_1$, $V - v_{20} \cos \theta_0 = v_{20} \sin \theta_0 \cot \theta_2$. From these two equations we must eliminate θ_0. To do so, we first solve for $\cos \theta_0$ and $\sin \theta_0$, and then form the sum of their squares, which is unity. Since $v_{10}/v_{20} = m_2/m_1$, we have finally, using (16.2),

$$(m_2/m_1) \sin^2\theta_2 + (m_1/m_2) \sin^2\theta_1 - 2 \sin\theta_1 \sin\theta_2 \cos(\theta_1+\theta_2)$$

$$= \frac{2\epsilon}{(m_1+m_2)V^2} \sin^2(\theta_1+\theta_2).$$

PROBLEM 2. Find the angular distribution of the resulting particles in the *L* system.

SOLUTION. When $v_0 > V$, we substitute (16.6), with the plus sign of the radical, in (16.7), obtaining

$$\tfrac{1}{2} \sin\theta \, d\theta \left[2\frac{V}{v_0} \cos\theta + \frac{1+(V^2/v_0^2) \cos 2\theta}{\sqrt{[1-(V^2/v_0^2) \sin^2\theta]}} \right] \quad (0 \leqslant \theta \leqslant \pi).$$

When $v_0 < V$, both possible relations between θ_0 and θ must be taken into account. Since, when θ increases, one value of θ_0 increases and the other decreases, the difference (not the sum) of the expressions d cos θ_0 with the two signs of the radical in (16.6) must be taken. The result is

$$\sin\theta \, d\theta \frac{1+(V^2/v_0^2) \cos 2\theta}{\sqrt{[1-(V^2/v_0^2) \sin^2\theta]}} \quad (0 \leqslant \theta \leqslant \theta_{max}).$$

PROBLEM 3. Determine the range of possible values of the angle θ between the directions of motion of the two resulting particles in the *L* system.

SOLUTION. The angle $\theta = \theta_1+\theta_2$, where θ_1 and θ_2 are the angles defined by formula (16.5) (see Problem 1), and it is simplest to calculate the tangent of θ. A consideration of the extrema of the resulting expression gives the following ranges of θ, depending on the relative magnitudes of V, v_{10} and v_{20} (for definiteness, we assume $v_{20} > v_{10}$): $0 < \theta < \pi$ if $v_{10} < V < v_{20}$, $\pi-\theta_0 < \theta < \pi$ if $V < v_{10}$, $0 < \theta < \theta_0$ if $V > v_{20}$. The value of θ_0 is given by

$$\sin\theta_0 = V(v_{10}+v_{20})/(V^2+v_{10}v_{20}).$$

§17. Elastic collisions

A collision between two particles is said to be *elastic* if it involves no change in their internal state. Accordingly, when the law of conservation of energy is applied to such a collision, the internal energy of the particles may be neglected.

The collision is most simply described in a frame of reference in which the centre of mass of the two particles is at rest (the *C* system). As in §16, we distinguish by the suffix 0 the values of quantities in that system. The velocities of the particles before the collision are related to their velocities $\mathbf{v}_1$ and $\mathbf{v}_2$ in the laboratory system by $\mathbf{v}_{10} = m_2\mathbf{v}/(m_1+m_2)$, $\mathbf{v}_{20} = -m_1\mathbf{v}/(m_1+m_2)$, where $\mathbf{v} = \mathbf{v}_1-\mathbf{v}_2$; see (13.2).

Because of the law of conservation of momentum, the momenta of the two particles remain equal and opposite after the collision, and are also unchanged in magnitude, by the law of conservation of energy. Thus, in the *C* system the collision simply rotates the velocities, which remain opposite in direction and unchanged in magnitude. If we denote by $\mathbf{n}_0$ a unit vector in the direction of the velocity of the particle m_1 after the collision, then the velocities of the two particles after the collision (distinguished by primes) are

$$\mathbf{v}_{10}' = m_2 v \mathbf{n}_0/(m_1+m_2), \qquad \mathbf{v}_{20}' = -m_1 v \mathbf{n}_0/(m_1+m_2). \tag{17.1}$$

In order to return to the L system, we must add to these expressions the velocity $\mathbf{V}$ of the centre of mass. The velocities in the L system after the collision are therefore

$$\mathbf{v_1}' = m_2 v \mathbf{n_0}/(m_1+m_2)+(m_1\mathbf{v_1}+m_2\mathbf{v_2})/(m_1+m_2),$$
$$\mathbf{v_2}' = -m_1 v \mathbf{n_0}/(m_1+m_2)+(m_1\mathbf{v_1}+m_2\mathbf{v_2})/(m_1+m_2).$$

(17.2)

No further information about the collision can be obtained from the laws of conservation of momentum and energy. The direction of the vector $\mathbf{n_0}$ depends on the law of interaction of the particles and on their relative position during the collision.

The results obtained above may be interpreted geometrically. Here it is more convenient to use momenta instead of velocities. Multiplying equations (17.2) by m_1 and m_2 respectively, we obtain

$$\mathbf{p_1}' = mv\mathbf{n_0}+m_1(\mathbf{p_1}+\mathbf{p_2})/(m_1+m_2),$$
$$\mathbf{p_2}' = -mv\mathbf{n_0}+m_2(\mathbf{p_1}+\mathbf{p_2})/(m_1+m_2),$$

(17.3)

where $m = m_1 m_2/(m_1+m_2)$ is the reduced mass. We draw a circle of radius mv and use the construction shown in Fig. 15. If the unit vector $\mathbf{n_0}$ is along OC, the vectors AC and CB give the momenta $\mathbf{p_1}'$ and $\mathbf{p_2}'$ respectively. When $\mathbf{p_1}$ and $\mathbf{p_2}$ are given, the radius of the circle and the points A and B are fixed, but the point C may be anywhere on the circle.

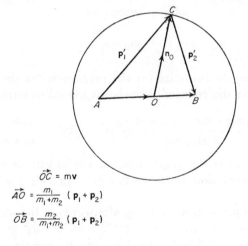

$$\vec{OC} = mv$$
$$\vec{AO} = \frac{m_1}{m_1+m_2}(\mathbf{P_1}+\mathbf{P_2})$$
$$\vec{OB} = \frac{m_2}{m_1+m_2}(\mathbf{P_1}+\mathbf{P_2})$$

Fig. 15

Let us consider in more detail the case where one of the particles (m_2, say) is at rest before the collision. In that case the distance $OB = m_2 p_1/(m_1+m_2) = mv$ is equal to the radius, i.e. B lies on the circle. The vector AB is equal to the momentum $\mathbf{p_1}$ of the particle m_1 before the collision. The point A lies inside or outside the circle, according as $m_1 < m_2$ or $m_1 > m_2$. The corresponding diagrams are shown in Figs. 16a, b. The angles θ_1 and θ_2 in these diagrams are the angles between the directions of motion after the collision and the direction of impact (i.e. of $\mathbf{p}_1$). The angle at the centre, denoted by χ, which gives the direction of $\mathbf{n}_0$, is the angle through which the direction of motion of m_1 is turned in the C system. It is evident from the figure that θ_1 and θ_2 can be expressed in terms of χ by

$$\tan\theta_1 = \frac{m_2 \sin\chi}{m_1 + m_2 \cos\chi}, \qquad \theta_2 = \tfrac{1}{2}(\pi-\chi). \qquad (17.4)$$

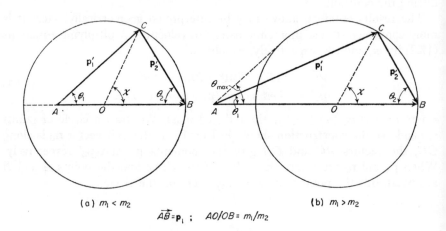

(a) $m_1 < m_2$ (b) $m_1 > m_2$

$$\overrightarrow{AB} = \mathbf{p}_1 \ ; \quad AO/OB = m_1/m_2$$

Fig. 16

We may give also the formulae for the magnitudes of the velocities of the two particles after the collision, likewise expressed in terms of χ:

$$v_1' = \frac{\sqrt{(m_1^2 + m_2^2 + 2m_1 m_2 \cos\chi)}}{m_1 + m_2}v, \qquad v_2' = \frac{2m_1 v}{m_1 + m_2}\sin\tfrac{1}{2}\chi. \qquad (17.5)$$

The sum $\theta_1 + \theta_2$ is the angle between the directions of motion of the particles after the collision. Evidently $\theta_1 + \theta_2 > \tfrac{1}{2}\pi$ if $m_1 < m_2$, and $\theta_1 + \theta_2 < \tfrac{1}{2}\pi$ if $m_1 > m_2$.

When the two particles are moving afterwards in the same or in opposite directions (head-on collision), we have $\chi = \pi$, i.e. the point C lies on the diameter through A, and is on OA (Fig. 16b; $\mathbf{p}_1'$ and $\mathbf{p}_2'$ in the same direction) or on OA produced (Fig. 16a; $\mathbf{p}_1'$ and $\mathbf{p}_2'$ in opposite directions).

In this case the velocities after the collision are

$$\mathbf{v}_1' = \frac{m_1 - m_2}{m_1 + m_2}\mathbf{v}, \qquad \mathbf{v}_2' = \frac{2m_1}{m_1 + m_2}\mathbf{v}. \qquad (17.6)$$

This value of $\mathbf{v}_2'$ has the greatest possible magnitude, and the maximum energy which can be acquired in the collision by a particle originally at rest is therefore

$$E_2'{}_{\max} = \tfrac{1}{2}m_2 v_2'{}^2{}_{\max} = \frac{4m_1 m_2}{(m_1 + m_2)^2} E_1, \tag{17.7}$$

where $E_1 = \tfrac{1}{2}m_1 v_1^2$ is the initial energy of the incident particle.

If $m_1 < m_2$, the velocity of m_1 after the collision can have any direction. If $m_1 > m_2$, however, this particle can be deflected only through an angle not exceeding $\theta_{\max}$ from its original direction; this maximum value of θ_1 corresponds to the position of C for which AC is a tangent to the circle (Fig. 16b). Evidently

$$\sin\theta_{\max} = OC/OA = m_2/m_1. \tag{17.8}$$

The collision of two particles of equal mass, of which one is initially at rest, is especially simple. In this case both B and A lie on the circle (Fig. 17).

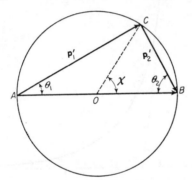

FIG. 17

Then

$$\theta_1 = \tfrac{1}{2}\chi, \qquad\qquad \theta_2 = \tfrac{1}{2}(\pi - \chi), \tag{17.9}$$

$$v_1' = v\cos\tfrac{1}{2}\chi, \qquad v_2' = v\sin\tfrac{1}{2}\chi. \tag{17.10}$$

After the collision the particles move at right angles to each other.

PROBLEM

Express the velocity of each particle after a collision between a moving particle (m_1) and another at rest (m_2) in terms of their directions of motion in the L system.

SOLUTION. From Fig. 16 we have $p_2' = 2OB\cos\theta_2$ or $v_2' = 2v(m/m_2)\cos\theta_2$. The momentum $p_1' = AC$ is given by $OC^2 = AO^2 + p_1'{}^2 - 2AO \cdot p_1'\cos\theta_1$ or

$$\left(\frac{v_1'}{v}\right)^2 - \frac{2m}{m_2}\frac{v_1'}{v}\cos\theta_1 + \frac{m_1 - m_2}{m_1 + m_2} = 0.$$

Hence

$$\frac{v_1'}{v} = \frac{m_1}{m_1 + m_2}\cos\theta_1 \pm \frac{1}{m_1 + m_2}\sqrt{(m_2^2 - m_1^2\sin^2\theta_1)};$$

for $m_1 > m_2$ the radical may have either sign, but for $m_2 > m_1$ it must be taken positive.

§18. Scattering

As already mentioned in §17, a complete calculation of the result of a collision between two particles (i.e. the determination of the angle χ) requires the solution of the equations of motion for the particular law of interaction involved.

We shall first consider the equivalent problem of the deflection of a single particle of mass m moving in a field $U(r)$ whose centre is at rest (and is at the centre of mass of the two particles in the original problem).

As has been shown in §14, the path of a particle in a central field is symmetrical about a line from the centre to the nearest point in the orbit (OA in Fig. 18). Hence the two asymptotes to the orbit make equal angles (ϕ_0, say) with this line. The angle χ through which the particle is deflected as it passes the centre is seen from Fig. 18 to be

$$\chi = |\pi - 2\phi_0|. \tag{18.1}$$

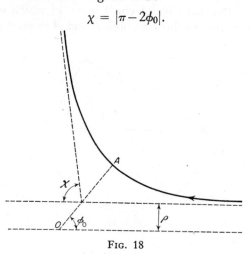

Fig. 18

The angle ϕ_0 itself is given, according to (14.7), by

$$\phi_0 = \int_{r_{min}}^{\infty} \frac{(M/r^2)\,dr}{\sqrt{\{2m[E - U(r)] - M^2/r^2\}}}, \tag{18.2}$$

taken between the nearest approach to the centre and infinity. It should be recalled that r_{min} is a zero of the radicand.

For an infinite motion, such as that considered here, it is convenient to use instead of the constants E and M the velocity v_∞ of the particle at infinity and the *impact parameter* ρ. The latter is the length of the perpendicular from the centre O to the direction of v_∞, i.e. the distance at which the particle would pass the centre if there were no field of force (Fig. 18). The energy and the angular momentum are given in terms of these quantities by

$$E = \tfrac{1}{2}mv_\infty^2, \qquad M = m\rho v_\infty, \tag{18.3}$$

and formula (18.2) becomes

$$\phi_0 = \int_{r_{min}}^{\infty} \frac{(\rho/r^2)\,dr}{\sqrt{[1-(\rho^2/r^2)-(2U/mv_\infty^2)]}}. \qquad (18.4)$$

Together with (18.1), this gives χ as a function of ρ.

In physical applications we are usually concerned not with the deflection of a single particle but with the *scattering* of a beam of identical particles incident with uniform velocity $\mathbf{v}_\infty$ on the scattering centre. The different particles in the beam have different impact parameters and are therefore scattered through different angles χ. Let dN be the number of particles scattered per unit time through angles between χ and $\chi+d\chi$. This number itself is not suitable for describing the scattering process, since it is proportional to the density of the incident beam. We therefore use the ratio

$$d\sigma = dN/n, \qquad (18.5)$$

where n is the number of particles passing in unit time through unit area of the beam cross-section (the beam being assumed uniform over its cross-section). This ratio has the dimensions of area and is called the *effective scattering cross-section*. It is entirely determined by the form of the scattering field and is the most important characteristic of the scattering process.

We shall suppose that the relation between χ and ρ is one-to-one; this is so if the angle of scattering is a monotonically decreasing function of the impact parameter. In that case, only those particles whose impact parameters lie between $\rho(\chi)$ and $\rho(\chi)+d\rho(\chi)$ are scattered at angles between χ and $\chi+d\chi$. The number of such particles is equal to the product of n and the area between two circles of radii ρ and $\rho+d\rho$, i.e. $dN = 2\pi\rho\,d\rho\,.\,n$. The effective cross-section is therefore

$$d\sigma = 2\pi\rho\,d\rho. \qquad (18.6)$$

In order to find the dependence of $d\sigma$ on the angle of scattering, we need only rewrite (18.6) as

$$d\sigma = 2\pi\rho(\chi)|d\rho(\chi)/d\chi|\,d\chi. \qquad (18.7)$$

Here we use the modulus of the derivative $d\rho/d\chi$, since the derivative may be (and usually is) negative.† Often $d\sigma$ is referred to the solid angle element do instead of the plane angle element $d\chi$. The solid angle between cones with vertical angles χ and $\chi+d\chi$ is $do = 2\pi\sin\chi\,d\chi$. Hence we have from (18.7)

$$d\sigma = \frac{\rho(\chi)}{\sin\chi}\left|\frac{d\rho}{d\chi}\right|\,do. \qquad (18.8)$$

---

† If the function $\rho(\chi)$ is many-valued, we must obviously take the sum of such expressions as (18.7) over all the branches of this function.

Returning now to the problem of the scattering of a beam of particles, not by a fixed centre of force, but by other particles initially at rest, we can say that (18.7) gives the effective cross-section as a function of the angle of scattering in the centre-of-mass system. To find the corresponding expression as a function of the scattering angle θ in the laboratory system, we must express χ in (18.7) in terms of θ by means of formulae (17.4). This gives expressions for both the scattering cross-section for the incident beam of particles (χ in terms of θ_1) and that for the particles initially at rest (χ in terms of θ_2).

PROBLEMS

PROBLEM 1. Determine the effective cross-section for scattering of particles from a perfectly rigid sphere of radius a (i.e. when the interaction is such that $U = \infty$ for $r < a$ and $U = 0$ for $r > a$).

SOLUTION. Since a particle moves freely outside the sphere and cannot penetrate into it, the path consists of two straight lines symmetrical about the radius to the point where the particle strikes the sphere (Fig. 19). It is evident from Fig. 19 that

$$\rho = a \sin \phi_0 = a \sin \tfrac{1}{2}(\pi - \chi) = a \cos \tfrac{1}{2}\chi.$$

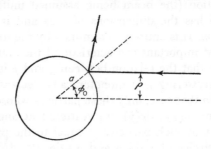

FIG. 19

Substituting in (18.7) or (18.8), we have

$$d\sigma = \tfrac{1}{2}\pi a^2 \sin \chi \, d\chi = \tfrac{1}{4}a^2 \, do, \tag{1}$$

i.e. the scattering is isotropic in the C system. On integrating $d\sigma$ over all angles, we find that the total cross-section $\sigma = \pi a^2$, in accordance with the fact that the "impact area" which the particle must strike in order to be scattered is simply the cross-sectional area of the sphere.

In order to change to the L system, χ must be expressed in terms of θ_1 by (17.4). The calculations are entirely similar to those of §16, Problem 2, on account of the formal resemblance between formulae (17.4) and (16.5). For $m_1 < m_2$ (where m_1 is the mass of the particle and m_2 that of the sphere) we have

$$d\sigma_1 = \tfrac{1}{4}a^2 \left[2(m_1/m_2) \cos \theta_1 + \frac{1 + (m_1/m_2)^2 \cos 2\theta_1}{\sqrt{[1 - (m_1/m_2)^2 \sin^2\theta_1]}} \right] do_1,$$

where $do_1 = 2\pi \sin \theta_1 \, d\theta_1$. If, on the other hand, $m_2 < m_1$, then

$$d\sigma_1 = \tfrac{1}{4}a^2 \frac{1 + (m_1/m_2)^2 \cos 2\theta_1}{\sqrt{[1 - (m_1/m_2)^2 \sin^2\theta]}} \, do_1.$$

For $m_1 = m_2$, we have $d\sigma_1 = a^2|\cos \theta_1| \, do_1$, which can also be obtained directly by substituting $\chi = 2\theta_1$ from (17.9) in (1).

For a sphere originally at rest, $\chi = \pi - 2\theta_2$ in all cases, and substitution in (1) gives

$$d\sigma_2 = a^2|\cos\theta_2|\,do_2.$$

PROBLEM 2. Express the effective cross-section (Problem 1) as a function of the energy ϵ lost by a scattered particle.

SOLUTION. The energy lost by a particle of mass m_1 is equal to that gained by the sphere of mass m_2. From (17.5) and (17.7), $\epsilon = E_2' = [2m_1^2 m_2/(m_1+m_2)^2]\,v_\infty^2\sin^2\tfrac{1}{2}\chi = \epsilon_{max}\sin^2\tfrac{1}{2}\chi$, whence $d\epsilon = \tfrac{1}{2}\epsilon_{max}\sin\chi\,d\chi$; substituting in (1), Problem 1, we have $d\sigma = \pi q^2\,d\epsilon/\epsilon_{max}$. The scattered particles are uniformly distributed with respect to ϵ in the range from zero to ϵ_{max}.

PROBLEM 3. Find the effective cross-section as a function of the velocity v_∞ for particles scattered in a field $U \sim r^{-n}$.

SOLUTION. According to (10.3), if the potential energy is a homogeneous function of order $k = -n$, then similar paths are such that $\rho \sim v^{-2/n}$, or $\rho = v_\infty^{-2/n}f(\chi)$, the angles of deflection χ being equal for similar paths. Substitution in (18.6) gives $d\sigma \sim v_\infty^{-4/n}\,do$.

PROBLEM 4. Determine the effective cross-section for a particle to "fall" to the centre of a field $U = -\alpha/r^2$.

SOLUTION. The particles which "fall" to the centre are those for which $2\alpha > m\rho^2 v_\infty^2$ (see (14.11)), i.e. for which the impact parameter does not exceed $\rho_{max} = \surd(2\alpha/mv_\infty^2)$. The effective cross-section is therefore $\sigma = \pi\rho_{max}^2 = 2\pi\alpha/mv_\infty^2$.

PROBLEM 5. The same as Problem 4, but for a field $U = -\alpha/r^n$ $(n > 2,\ \alpha > 0)$.

SOLUTION. The effective potential energy $U_{eff} = m\rho^2 v_\infty^2/2r^2 - \alpha/r^n$ depends on r in the manner shown in Fig. 20. Its maximum value is

$$U_{eff,max} \equiv U_0 = \tfrac{1}{2}(n-2)\alpha(m\rho^2 v_\infty^2/\alpha n)^{n/(n-2)}.$$

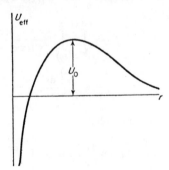

FIG. 20

The particles which "fall" to the centre are those for which $U_0 < E$. The condition $U_0 = E$ gives ρ_{max}, whence

$$\sigma = \pi n(n-2)^{(2-n)/n}(\alpha/mv_\infty^2)^{2/n}.$$

PROBLEM 6. Determine the effective cross-section for particles of mass m_1 to strike a sphere of mass m_2 and radius R to which they are attracted in accordance with Newton's law.

SOLUTION. The condition for a particle to reach the sphere is that $r_{min} < R$, where r_{min} is the point on the path which is nearest to the centre of the sphere. The greatest possible value of ρ is given by $r_{min} = R$; this is equivalent to $U_{eff}(R) = E$ or $\tfrac{1}{2}m_1 v_\infty^2\rho_{max}^2/R^2 - \alpha/R = \tfrac{1}{2}m_1 v_\infty^2$, where $\alpha = \gamma m_1 m_2$ (γ being the gravitational constant) and we have put $m \approx m_1$ on the assumption that $m_2 \gg m_1$. Solving for ρ_{max}^2, we finally obtain $\sigma = \pi R^2(1 + 2\gamma m_2/Rv_\infty^2)$.

When $v_\infty \to \infty$ the effective cross-section tends, of course, to the geometrical cross-section of the sphere.

PROBLEM 7. Deduce the form of a scattering field $U(r)$, given the effective cross-section as a function of the angle of scattering for a given energy E. It is assumed that $U(r)$ decreases monotonically with r (a repulsive field), with $U(0) > E$ and $U(\infty) = 0$ (O. B. FIRSOV 1953).

SOLUTION. Integration of $d\sigma$ with respect to the scattering angle gives, according to the formula

$$\int_\chi^\pi (d\sigma/d\chi)\,d\chi = \pi\rho^2, \tag{1}$$

the square of the impact parameter, so that $\rho(\chi)$ (and therefore $\chi(\rho)$) is known.
We put

$$s = 1/r, \qquad x = 1/\rho^2, \qquad w = \sqrt{[1-(U/E)]}. \tag{2}$$

Then formulae (18.1), (18.2) become

$$\tfrac12[\pi - \chi(x)] = \int_0^{s_0} \frac{ds}{\sqrt{(xw^2 - s^2)}}, \tag{3}$$

where $s_0(x)$ is the root of the equation $xw^2(s_0) - s_0^2 = 0$.

Equation (3) is an integral equation for the function $w(s)$, and may be solved by a method similar to that used in §12. Dividing both sides of (3) by $\sqrt{(\alpha - x)}$ and integrating with respect to x from zero to α, we find

$$\int_0^\alpha \frac{\pi - \chi(x)}{2} \cdot \frac{dx}{\sqrt{(\alpha - x)}} = \int_0^\alpha \int_0^{s_0(x)} \frac{ds\,dx}{\sqrt{[(xw^2 - s^2)(\alpha - x)]}}$$

$$= \int_0^{s_0(\alpha)} \int_{x(s_0)}^\alpha \frac{dx\,ds}{\sqrt{[(xw^2 - s^2)(\alpha - x)]}}$$

$$= \pi \int_0^{s_0(\alpha)} \frac{ds}{w},$$

or, integrating by parts on the left-hand side,

$$\pi\sqrt{\alpha} - \int_0^\alpha \sqrt{(\alpha - x)}\frac{d\chi}{dx}\,dx = \pi \int_0^{s_0(\alpha)} \frac{ds}{w}.$$

This relation is differentiated with respect to α, and then $s_0(\alpha)$ is replaced by s simply; accordingly α is replaced by s^2/w^2, and the result is, in differential form,

$$\pi\,d(s/w) - \tfrac12 d(s^2/w^2)\int_0^{s^2/w^2} \frac{\chi'(x)\,dx}{\sqrt{[(s^2/w^2) - x]}} = (\pi/w)\,ds$$

or

$$-\pi\,d\log w = d(s/w)\int_0^{s^2/w^2} \frac{\chi'(x)\,dx}{\sqrt{[(s^2/w^2) - x]}}.$$

This equation can be integrated immediately if the order of integration on the right-hand side is inverted. Since for $s = 0$ (i.e. $r \to \infty$) we must have $w = 1$ (i.e. $U = 0$), we have, on returning to the original variables r and ρ, the following two equivalent forms of the final result:

$$w = \exp\left\{\frac{1}{\pi}\int\limits_{rw}^{\infty} \cosh^{-1}(\rho/rw)\,(\mathrm{d}\chi/\mathrm{d}\rho)\,\mathrm{d}\rho\right\}$$

$$= \exp\left\{-\frac{1}{\pi}\int\limits_{rw}^{\infty} \frac{\chi(\rho)\,\mathrm{d}\rho}{\sqrt{(\rho^2 - r^2 w^2)}}\right\}. \tag{4}$$

This formula determines implicitly the function $w(r)$ (and therefore $U(r)$) for all $r > r_{\min}$, i.e. in the range of r which can be reached by a scattered particle of given energy E.

§19. Rutherford's formula

One of the most important applications of the formulae derived above is to the scattering of charged particles in a Coulomb field. Putting in (18.4) $U = \alpha/r$ and effecting the elementary integration, we obtain

$$\phi_0 = \cos^{-1}\frac{\alpha/mv_\infty^2 \rho}{\sqrt{[1 + (\alpha/mv_\infty^2 \rho)^2]}},$$

whence $\rho^2 = (\alpha^2/m^2 v_\infty^4)\tan^2\phi_0$, or, putting $\phi_0 = \frac{1}{2}(\pi - \chi)$ from (18.1),

$$\rho^2 = (\alpha^2/m^2 v_\infty^4)\cot^2\tfrac{1}{2}\chi. \tag{19.1}$$

Differentiating this expression with respect to χ and substituting in (18.7) or (18.8) gives

$$\mathrm{d}\sigma = \pi(\alpha/mv_\infty^2)^2 \cos\tfrac{1}{2}\chi\,\mathrm{d}\chi/\sin^3\tfrac{1}{2}\chi \tag{19.2}$$

or

$$\mathrm{d}\sigma = (\alpha/2mv_\infty^2)^2\,\mathrm{d}o/\sin^4\tfrac{1}{2}\chi. \tag{19.3}$$

This is *Rutherford's formula*. It may be noted that the effective cross-section is independent of the sign of α, so that the result is equally valid for repulsive and attractive Coulomb fields.

Formula (19.3) gives the effective cross-section in the frame of reference in which the centre of mass of the colliding particles is at rest. The transformation to the laboratory system is effected by means of formulae (17.4). For particles initially at rest we substitute $\chi = \pi - 2\theta_2$ in (19.2) and obtain

$$\mathrm{d}\sigma_2 = 2\pi(\alpha/mv_\infty^2)^2 \sin\theta_2\,\mathrm{d}\theta_2/\cos^3\theta_2$$

$$= (\alpha/mv_\infty^2)^2\,\mathrm{d}o_2/\cos^3\theta_2. \tag{19.4}$$

The same transformation for the incident particles leads, in general, to a very complex formula, and we shall merely note two particular cases.

If the mass m_2 of the scattering particle is large compared with the mass m_1 of the scattered particle, then $\chi \approx \theta_1$ and $m \approx m_1$, so that

$$\mathrm{d}\sigma_1 = (\alpha/4E_1)^2\,\mathrm{d}o_1/\sin^4\tfrac{1}{2}\theta_1, \tag{19.5}$$

where $E_1 = \frac{1}{2}m_1 v_\infty^2$ is the energy of the incident particle.

If the masses of the two particles are equal ($m_1 = m_2$, $m = \frac{1}{2}m_1$), then by (17.9) $\chi = 2\theta_1$, and substitution in (19.2) gives

$$d\sigma_1 = 2\pi(\alpha/E_1)^2 \cos\theta_1 \, d\theta_1/\sin^3\theta_1$$
$$= (\alpha/E_1)^2 \cos\theta_1 \, do_1/\sin^4\theta_1. \tag{19.6}$$

If the particles are entirely identical, that which was initially at rest cannot be distinguished after the collision. The total effective cross-section for all particles is obtained by adding $d\sigma_1$ and $d\sigma_2$, and replacing θ_1 and θ_2 by their common value θ:

$$d\sigma = (\alpha/E_1)^2\left(\frac{1}{\sin^4\theta} + \frac{1}{\cos^4\theta}\right)\cos\theta \, do. \tag{19.7}$$

Let us return to the general formula (19.2) and use it to determine the distribution of the scattered particles with respect to the energy lost in the collision. When the masses of the scattered (m_1) and scattering (m_2) particles are arbitrary, the velocity acquired by the latter is given in terms of the angle of scattering in the C system by $v_2' = [2m_1/(m_1+m_2)]v_\infty \sin\frac{1}{2}\chi$; see (17.5). The energy acquired by m_2 and lost by m_1 is therefore $\epsilon = \frac{1}{2}m_2v_2'^2 = (2m^2/m_2)v_\infty^2 \sin^2\frac{1}{2}\chi$. Expressing $\sin\frac{1}{2}\chi$ in terms of ϵ and substituting in (19.2), we obtain

$$d\sigma = 2\pi(\alpha^2/m_2v_\infty^2) \, d\epsilon/\epsilon^2. \tag{19.8}$$

This is the required formula: it gives the effective cross-section as a function of the energy loss ϵ, which takes values from zero to $\epsilon_{\max} = 2m^2v_\infty^2/m_2$.

PROBLEMS

PROBLEM 1. Find the effective cross-section for scattering in a field $U = \alpha/r^2$ ($\alpha > 0$).

SOLUTION. The angle of deflection is

$$\chi = \pi\left[1 - \frac{1}{\sqrt{\{1+2\alpha/mp^2v_\infty^2\}}}\right].$$

The effective cross-section is

$$d\sigma = \frac{2\pi^2\alpha}{mv_\infty^2} \cdot \frac{\pi-\chi}{\chi^2(2\pi-\chi)^2} \cdot \frac{do}{\sin\chi}.$$

PROBLEM 2. Find the effective cross-section for scattering by a spherical "potential well" of radius a and "depth" U_0 (i.e. a field with $U = 0$ for $r > a$ and $U = -U_0$ for $r < a$).

SOLUTION. The particle moves in a straight line which is "refracted" on entering and leaving the well. According to §7, Problem, the angle of incidence α and the angle of refraction β (Fig. 21) are such that $\sin\alpha/\sin\beta = n$, where $n = \sqrt{(1+2U_0/mv_\infty^2)}$. The angle of deflection is $\chi = 2(\alpha-\beta)$. Hence

$$\frac{\sin(\alpha-\frac{1}{2}\chi)}{\sin\alpha} = \cos\frac{1}{2}\chi - \cot\alpha\sin\frac{1}{2}\chi = \frac{1}{n}.$$

Eliminating α from this equation and the relation $a\sin\alpha = \rho$, which is evident from the diagram, we find the relation between ρ and χ:

$$\rho^2 = a^2\frac{n^2\sin^2\frac{1}{2}\chi}{n^2+1-2n\cos\frac{1}{2}\chi}.$$

Finally, differentiating, we have the effective cross-section:

$$d\sigma = \frac{a^2 n^2}{4 \cos \tfrac{1}{2}\chi} \frac{(n \cos \tfrac{1}{2}\chi - 1)(n - \cos \tfrac{1}{2}\chi)}{(n^2 + 1 - 2n \cos \tfrac{1}{2}\chi)^2} do.$$

The angle χ varies from zero (for $\rho = 0$) to χ_{max} (for $\rho = a$), where $\cos \tfrac{1}{2}\chi_{max} = 1/n$.

The total effective cross-section, obtained by integrating $d\sigma$ over all angles within the cone $\chi < \chi_{max}$, is, of course, equal to the geometrical cross-section πa^2.

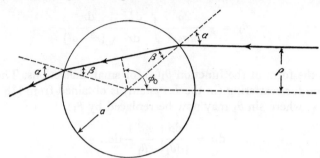

FIG. 21

§20. Small-angle scattering

The calculation of the effective cross-section is much simplified if only those collisions are considered for which the impact parameter is large, so that the field U is weak and the angles of deflection are small. The calculation can be carried out in the laboratory system, and the centre-of-mass system need not be used.

We take the x-axis in the direction of the initial momentum of the scattered particle m_1, and the xy-plane in the plane of scattering. Denoting by $\mathbf{p_1}'$ the momentum of the particle after scattering, we evidently have $\sin \theta_1 = p_{1y}'/p_1'$. For small deflections, $\sin \theta_1$ may be approximately replaced by θ_1, and p_1' in the denominator by the initial momentum $p_1 = m_1 v_\infty$:

$$\theta_1 \approx p_{1y}'/m_1 v_\infty. \tag{20.1}$$

Next, since $\dot{p}_y = F_y$, the total increment of momentum in the y-direction is

$$p_{1y}' = \int_{-\infty}^{\infty} F_y \, dt. \tag{20.2}$$

The force $F_y = -\partial U/\partial y = -(dU/dr)\partial r/\partial y = -(dU/dr)y/r$.

Since the integral (20.2) already contains the small quantity U, it can be calculated, in the same approximation, by assuming that the particle is not deflected at all from its initial path, i.e. that it moves in a straight line $y = \rho$ with uniform velocity v_∞. Thus we put in (20.2) $F_y = -(dU/dr)\rho/r$, $dt = dx/v_\infty$. The result is

$$p_{1y}' = -\frac{\rho}{v_\infty} \int_{-\infty}^{\infty} \frac{dU}{dr} \frac{dx}{r}.$$

Finally, we change the integration over x to one over r. Since, for a straight path, $r^2 = x^2 + \rho^2$, when x varies from $-\infty$ to $+\infty$, r varies from ∞ to ρ and back. The integral over x therefore becomes twice the integral over r from ρ to ∞, and $dx = r\, dr/\sqrt{(r^2 - \rho^2)}$. The angle of scattering θ_1 is thus given by†

$$\theta_1 = -\frac{2\rho}{m_1 v_\infty{}^2} \int_\rho^\infty \frac{dU}{dr} \frac{dr}{\sqrt{(r^2 - \rho^2)}}, \tag{20.3}$$

and this is the form of the function $\theta_1(\rho)$ for small deflections. The effective cross-section for scattering (in the L system) is obtained from (18.8) with θ_1 instead of χ, where $\sin \theta_1$ may now be replaced by θ_1:

$$d\sigma = \left| \frac{d\rho}{d\theta_1} \right| \frac{\rho(\theta_1)}{\theta_1} \, do_1. \tag{20.4}$$

PROBLEMS

PROBLEM 1. Derive formula (20.3) from (18.4).

SOLUTION. In order to avoid spurious divergences, we write (18.4) in the form

$$\phi_0 = -\frac{\partial}{\partial \rho} \int_{r_{\min}}^R \sqrt{\left[1 - \frac{\rho^2}{r^2} - \frac{2U}{mv_\infty{}^2}\right]} dr,$$

and take as the upper limit some large finite quantity R, afterwards taking the value as $R \to \infty$. Since U is small, we expand the square root in powers of U, and approximately replace $r_{\min}$ by ρ:

$$\phi_0 = \int_\rho^R \frac{\rho \, dr}{r^2 \sqrt{(1 - \rho^2/r^2)}} + \frac{\partial}{\partial \rho} \int_\rho^\infty \frac{U(r) \, dr}{mv_\infty{}^2 \sqrt{(1 - \rho^2/r^2)}}.$$

The first integral tends to $\tfrac{1}{2}\pi$ as $R \to \infty$. The second integral is integrated by parts, giving

$$= \pi - 2\phi_0 = 2\frac{\partial}{\partial \rho} \int_\rho^\infty \frac{\sqrt{(r^2 - \rho^2)}}{mv_\infty{}^2} \frac{dU}{dr} dr$$

$$= -\frac{2\rho}{mv_\infty{}^2} \int_\rho^\infty \frac{dU}{dr} \frac{dr}{\sqrt{(r^2 - \rho^2)}}.$$

This is equivalent to (20.3).

PROBLEM 2. Determine the effective cross-section for small-angle scattering in a field $U = \alpha/r^n$ $(n > 0)$.

---

† If the above derivation is applied in the C system, the expression obtained for χ is the same with m in place of m_1, in accordance with the fact that the small angles θ_1 and χ are related by (see (17.4)) $\theta_1 = m_2\chi/(m_1 + m_2)$.

SOLUTION. From (20.3) we have

$$\theta_1 = \frac{2\rho\alpha n}{m_1 v_\infty^2} \int\limits_{\rho}^{\infty} \frac{dr}{r^{n+1}\sqrt{(r^2-\rho^2)}}.$$

The substitution $\rho^2/r^2 = u$ converts the integral to a beta function, which can be expressed in terms of gamma functions:

$$\theta_1 = \frac{2\alpha\sqrt{\pi}}{m_1 v_\infty^2 \rho^n} \cdot \frac{\Gamma(\frac{1}{2}n+\frac{1}{2})}{\Gamma(\frac{1}{2}n)}.$$

Expressing ρ in terms of θ_1 and substituting in (20.4), we obtain

$$d\sigma = \frac{1}{n}\left[\frac{2\sqrt{\pi}\Gamma(\frac{1}{2}n+\frac{1}{2})}{\Gamma(\frac{1}{2}n)} \cdot \frac{\alpha}{m_1 v_\infty^2}\right]^{2/n} \theta_1^{-2-2/n} \, do_1.$$

CHAPTER V

SMALL OSCILLATIONS

§21. Free oscillations in one dimension

A VERY common form of motion of mechanical systems is what are called *small oscillations* of a system about a position of stable equilibrium. We shall consider first of all the simplest case, that of a system with only one degree of freedom.

Stable equilibrium corresponds to a position of the system in which its potential energy $U(q)$ is a minimum. A movement away from this position results in the setting up of a force $-dU/dq$ which tends to return the system to equilibrium. Let the equilibrium value of the generalised co-ordinate q be q_0. For small deviations from the equilibrium position, it is sufficient to retain the first non-vanishing term in the expansion of the difference $U(q) - U(q_0)$ in powers of $q - q_0$. In general this is the second-order term: $U(q) - U(q_0) \cong \frac{1}{2}k(q - q_0)^2$, where k is a positive coefficient, the value of the second derivative $U''(q)$ for $q = q_0$. We shall measure the potential energy from its minimum value, i.e. put $U(q_0) = 0$, and use the symbol

$$x = q - q_0 \tag{21.1}$$

for the deviation of the co-ordinate from its equilibrium value. Thus

$$U(x) = \tfrac{1}{2}kx^2. \tag{21.2}$$

The kinetic energy of a system with one degree of freedom is in general of the form $\frac{1}{2}a(q)\dot{q}^2 = \frac{1}{2}a(q)\dot{x}^2$. In the same approximation, it is sufficient to replace the function $a(q)$ by its value at $q = q_0$. Putting for brevity† $a(q_0) = m$, we have the following expression for the Lagrangian of a system executing small oscillations in one dimension:‡

$$L = \tfrac{1}{2}m\dot{x}^2 - \tfrac{1}{2}kx^2. \tag{21.3}$$

The corresponding equation of motion is

$$m\ddot{x} + kx = 0, \tag{21.4}$$

or

$$\ddot{x} + \omega^2 x = 0, \tag{21.5}$$

where

$$\omega = \sqrt{(k/m)}. \tag{21.6}$$

---

† It should be noticed that m is the mass only if x is the Cartesian co-ordinate.
‡ Such a system is often called a *one-dimensional oscillator*.

Two independent solutions of the linear differential equation (21.5) are $\cos \omega t$ and $\sin \omega t$, and its general solution is therefore

$$x = c_1 \cos \omega t + c_2 \sin \omega t. \tag{21.7}$$

This expression can also be written

$$x = a \cos(\omega t + \alpha). \tag{21.8}$$

Since $\cos(\omega t + \alpha) = \cos \omega t \cos \alpha - \sin \omega t \sin \alpha$, a comparison with (21.7) shows that the arbitrary constants a and α are related to c_1 and c_2 by

$$a = \sqrt{(c_1^2 + c_2^2)}, \qquad \tan \alpha = -c_2/c_1. \tag{21.9}$$

Thus, near a position of stable equilibrium, a system executes harmonic oscillations. The coefficient a of the periodic factor in (21.8) is called the *amplitude* of the oscillations, and the argument of the cosine is their *phase*; α is the initial value of the phase, and evidently depends on the choice of the origin of time. The quantity ω is called the *angular frequency* of the oscillations; in theoretical physics, however, it is usually called simply the *frequency*, and we shall use this name henceforward.

The frequency is a fundamental characteristic of the oscillations, and is independent of the initial conditions of the motion. According to formula (21.6) it is entirely determined by the properties of the mechanical system itself. It should be emphasised, however, that this property of the frequency depends on the assumption that the oscillations are small, and ceases to hold in higher approximations. Mathematically, it depends on the fact that the potential energy is a quadratic function of the co-ordinate.†

The energy of a system executing small oscillations is $E = \tfrac{1}{2}m\dot{x}^2 + \tfrac{1}{2}kx^2 = \tfrac{1}{2}m(\dot{x}^2 + \omega^2 x^2)$ or, substituting (21.8),

$$E = \tfrac{1}{2}m\omega^2 a^2. \tag{21.10}$$

It is proportional to the square of the amplitude.

The time dependence of the co-ordinate of an oscillating system is often conveniently represented as the real part of a complex expression:

$$x = \mathrm{re}\,[A \exp(i\omega t)], \tag{21.11}$$

where A is a complex constant; putting

$$A = a \exp(i\alpha), \tag{21.12}$$

we return to the expression (21.8). The constant A is called the *complex amplitude*; its modulus is the ordinary amplitude, and its argument is the initial phase.

The use of exponential factors is mathematically simpler than that of trigonometrical ones because they are unchanged in form by differentiation.

---

† It therefore does not hold good if the function $U(x)$ has at $x = 0$ a minimum of higher order, i.e. $U \sim x^n$ with $n > 2$; see §11, Problem 2(a).

So long as all the operations concerned are linear (addition, multiplication by constants, differentiation, integration), we may omit the sign re throughout and take the real part of the final result.

PROBLEMS

PROBLEM 1. Express the amplitude and initial phase of the oscillations in terms of the initial co-ordinate x_0 and velocity v_0.

SOLUTION. $a = \sqrt{(x_0^2 + v_0^2/\omega^2)}$, $\tan \alpha = -v_0/\omega x_0$.

PROBLEM 2. Find the ratio of frequencies ω and ω' of the oscillations of two diatomic molecules consisting of atoms of different isotopes, the masses of the atoms being m_1, m_2 and m_1', m_2'.

SOLUTION. Since the atoms of the isotopes interact in the same way, we have $k = k'$. The coefficients m in the kinetic energies of the molecules are their reduced masses. According to (21.6) we therefore have

$$\frac{\omega'}{\omega} = \sqrt{\frac{m_1 m_2 (m_1' + m_2')}{m_1' m_2' (m_1 + m_2)}}.$$

PROBLEM 3. Find the frequency of oscillations of a particle of mass m which is free to move along a line and is attached to a spring whose other end is fixed at a point A (Fig. 22) at a distance l from the line. A force F is required to extend the spring to length l.

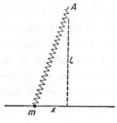

FIG. 22

SOLUTION. The potential energy of the spring is (to within higher-order terms) equal to the force F multiplied by the extension δl of the spring. For $x \ll l$ we have $\delta l = \sqrt{(l^2 + x^2)} - l = x^2/2l$, so that $U = Fx^2/2l$. Since the kinetic energy is $\frac{1}{2}m\dot{x}^2$, we have $\omega = \sqrt{(F/ml)}$.

PROBLEM 4. The same as Problem 3, but for a particle of mass m moving on a circle of radius r (Fig. 23).

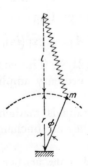

FIG. 23

SOLUTION. In this case the extension of the spring is (if $\phi \ll 1$)

$$\delta l = \sqrt{[r^2+(l+r)^2-2r(l+r) \cos \phi]} - l \approx r(l+r)\phi^2/2l.$$

The kinetic energy is $T = \tfrac{1}{2}mr^2\dot\phi^2$, and the frequency is therefore $\omega = \sqrt{[F(r+l)/mrl]}$.

PROBLEM 5. Find the frequency of oscillations of the pendulum shown in Fig. 2 (§5), whose point of support carries a mass m_1 and is free to move horizontally.

SOLUTION. For $\phi \ll 1$ the formula derived in §14, Problem 3, gives

$$T = \tfrac{1}{2}m_1m_2l^2\dot\phi^2/(m_1+m_2), \qquad U = \tfrac{1}{2}m_2gl\phi^2.$$

Hence

$$\omega = \sqrt{[g(m_1+m_2)/m_1l]}.$$

PROBLEM 6. Determine the form of a curve such that the frequency of oscillations of a particle on it under the force of gravity is independent of the amplitude.

SOLUTION. The curve satisfying the given condition is one for which the potential energy of a particle moving on it is $U = \tfrac{1}{2}ks^2$, where s is the length of the arc from the position of equilibrium. The kinetic energy $T = \tfrac{1}{2}m\dot s^2$, where m is the mass of the particle, and the frequency is then $\omega = \sqrt{(k/m)}$ whatever the initial value of s.

In a gravitational field $U = mgy$, where y is the vertical co-ordinate. Hence we have $\tfrac{1}{2}ks^2 = mgy$ or $y = \omega^2s^2/2g$. But $ds^2 = dx^2+dy^2$, whence

$$x = \int\sqrt{[(ds/dy)^2-1]} \, dy = \int\sqrt{[(g/2\omega^2y)-1]} \, dy.$$

The integration is conveniently effected by means of the substitution $y = g(1-\cos \xi)/4\omega^2$, which yields $x = g(\xi+\sin \xi)/4\omega^2$. These two equations give, in parametric form, the equation of the required curve, which is a cycloid.

§22. Forced oscillations

Let us now consider oscillations of a system on which a variable external force acts. These are called *forced* oscillations, whereas those discussed in §21 are *free* oscillations. Since the oscillations are again supposed small, it is implied that the external field is weak, because otherwise it could cause the displacement x to take too large values.

The system now has, besides the potential energy $\tfrac{1}{2}kx^2$, the additional potential energy $U_e(x, t)$ resulting from the external field. Expanding this additional term as a series of powers of the small quantity x, we have $U_e(x, t) \cong U_e(0, t)+x[\partial U_e/\partial x]_{x=0}$. The first term is a function of time only, and may therefore be omitted from the Lagrangian, as being the total time derivative of another function of time. In the second term $-[\partial U_e/\partial x]_{x=0}$ is the external "force" acting on the system in the equilibrium position, and is a given function of time, which we denote by $F(t)$. Thus the potential energy involves a further term $-xF(t)$, and the Lagrangian of the system is

$$L = \tfrac{1}{2}m\dot x^2-\tfrac{1}{2}kx^2+xF(t). \qquad (22.1)$$

The corresponding equation of motion is $m\ddot x+kx = F(t)$ or

$$\ddot x+\omega^2x = F(t)/m, \qquad (22.2)$$

where we have again introduced the frequency ω of the free oscillations.

The general solution of this inhomogeneous linear differential equation with constant coefficients is $x = x_0+x_1$, where x_0 is the general solution of the corresponding homogeneous equation and x_1 is a particular integral of the inhomogeneous equation. In the present case x_0 represents the free oscillations discussed in §21.

Let us consider a case of especial interest, where the external force is itself a simple periodic function of time, of some frequency γ:

$$F(t) = f \cos(\gamma t + \beta). \tag{22.3}$$

We seek a particular integral of equation (22.2) in the form $x_1 = b \cos(\gamma t + \beta)$, with the same periodic factor. Substitution in that equation gives $b = f/m(\omega^2 - \gamma^2)$; adding the solution of the homogeneous equation, we obtain the general integral in the form

$$x = a \cos(\omega t + \alpha) + [f/m(\omega^2 - \gamma^2)] \cos(\gamma t + \beta). \tag{22.4}$$

The arbitrary constants a and α are found from the initial conditions.

Thus a system under the action of a periodic force executes a motion which is a combination of two oscillations, one with the intrinsic frequency ω of the system and one with the frequency γ of the force.

The solution (22.4) is not valid when *resonance* occurs, i.e. when the frequency γ of the external force is equal to the intrinsic frequency ω of the system. To find the general solution of the equation of motion in this case, we rewrite (22.4) as

$$x = a \cos(\omega t + \alpha) + [f/m(\omega^2 - \gamma^2)][\cos(\gamma t + \beta) - \cos(\omega t + \beta)],$$

where a now has a different value. As $\gamma \to \omega$, the second term is indeterminate, of the form $0/0$. Resolving the indeterminacy by L'Hospital's rule, we have

$$x = a \cos(\omega t + \alpha) + (f/2m\omega) t \sin(\omega t + \beta). \tag{22.5}$$

Thus the amplitude of oscillations in resonance increases linearly with the time (until the oscillations are no longer small and the whole theory given above becomes invalid).

Let us also ascertain the nature of small oscillations near resonance, when $\gamma = \omega + \epsilon$ with ϵ a small quantity. We put the general solution in the complex form

$$x = A \exp(i\omega t) + B \exp[i(\omega + \epsilon)t] = [A + B \exp(i\epsilon t)] \exp(i\omega t). \tag{22.6}$$

Since the quantity $A + B \exp(i\epsilon t)$ varies only slightly over the period $2\pi/\omega$ of the factor $\exp(i\omega t)$, the motion near resonance may be regarded as small oscillations of variable amplitude.† Denoting this amplitude by C, we have $C = |A + B \exp(i\epsilon t)|$. Writing A and B in the form $a \exp(i\alpha)$ and $b \exp(i\beta)$ respectively, we obtain

$$C^2 = a^2 + b^2 + 2ab \cos(\epsilon t + \beta - \alpha). \tag{22.7}$$

---

† The "constant" term in the phase of the oscillation also varies.

Thus the amplitude varies periodically with frequency ϵ between the limits $|a-b| \leqslant C \leqslant a+b$. This phenomenon is called *beats*.

The equation of motion (22.2) can be integrated in a general form for an arbitrary external force $F(t)$. This is easily done by rewriting the equation as

$$\frac{d}{dt}(\dot{x}+i\omega x) - i\omega(\dot{x}+i\omega x) = \frac{1}{m}F(t)$$

or

$$d\xi/dt - i\omega\xi = F(t)/m, \qquad (22.8)$$

where

$$\xi = \dot{x} + i\omega x \qquad (22.9)$$

is a complex quantity. Equation (22.8) is of the first order. Its solution when the right-hand side is replaced by zero is $\xi = A \exp(i\omega t)$ with constant A. As before, we seek a solution of the inhomogeneous equation in the form $\xi = A(t) \exp(i\omega t)$, obtaining for the function $A(t)$ the equation $\dot{A}(t) = F(t) \exp(-i\omega t)/m$. Integration gives the solution of (22.9):

$$\xi = \exp(i\omega t)\left\{ \int_0^t \frac{1}{m}F(t)\exp(-i\omega t)\,dt + \xi_0 \right\}, \qquad (22.10)$$

where the constant of integration ξ_0 is the value of ξ at the instant $t = 0$. This is the required general solution; the function $x(t)$ is given by the imaginary part of (22.10), divided by ω.†

The energy of a system executing forced oscillations is naturally not conserved, since the system gains energy from the source of the external field. Let us determine the total energy transmitted to the system during all time, assuming its initial energy to be zero. According to formula (22.10), with the lower limit of integration $-\infty$ instead of zero and with $\xi(-\infty) = 0$, we have for $t \to \infty$

$$|\xi(\infty)|^2 = \frac{1}{m^2}\left| \int_{-\infty}^{\infty} F(t)\exp(-i\omega t)\,dt \right|^2.$$

The energy of the system is

$$E = \tfrac{1}{2}m(\dot{x}^2 + \omega^2 x^2) = \tfrac{1}{2}m|\xi|^2. \qquad (22.11)$$

Substituting $|\xi(\infty)|^2$, we obtain the energy transferred:

$$E = \frac{1}{2m}\left| \int_{-\infty}^{\infty} F(t)\exp(-i\omega t)\,dt \right|^2; \qquad (22.12)$$

---

† The force $F(t)$ must, of course, be written in real form.

it is determined by the squared modulus of the Fourier component of the force $F(t)$ whose frequency is the intrinsic frequency of the system.

In particular, if the external force acts only during a time short in comparison with $1/\omega$, we can put $\exp(-i\omega t) \cong 1$. Then

$$ E = \frac{1}{2m} \left(\int_{-\infty}^{\infty} F(t)\, dt \right)^2. $$

This result is obvious: it expresses the fact that a force of short duration gives the system a momentum $\int F\, dt$ without bringing about a perceptible displacement.

PROBLEMS

PROBLEM 1. Determine the forced oscillations of a system under a force $F(t)$ of the following forms, if at time $t = 0$ the system is at rest in equilibrium $(x = \dot{x} = 0)$: (a) $F = F_0$, a constant, (b) $F = at$, (c) $F = F_0 \exp(-\alpha t)$, (d) $F = F_0 \exp(-\alpha t) \cos \beta t$.

SOLUTION. (a) $x = (F_0/m\omega^2)(1 - \cos \omega t)$. The action of the constant force results in a displacement of the position of equilibrium about which the oscillations take place.

(b) $x = (a/m\omega^3)(\omega t - \sin \omega t)$.

(c) $x = [F_0/m(\omega^2 + \alpha^2)][\exp(-\alpha t) - \cos \omega t + (\alpha/\omega) \sin \omega t]$.

(d) $x = F_0\{-(\omega^2 + \alpha^2 - \beta^2) \cos \omega t + (\alpha/\omega)(\omega^2 + \alpha^2 + \beta^2) \sin \omega t +$
$+ \exp(-\alpha t)[(\omega^2 + \alpha^2 - \beta^2) \cos \beta t - 2\alpha\beta \sin \beta t]\}/m[(\omega^2 + \alpha^2 - \beta^2)^2 + 4\alpha^2\beta^2]$.

This last case is conveniently treated by writing the force in the complex form

$$ F = F_0 \exp[(-\alpha + i\beta)t]. $$

PROBLEM 2. Determine the final amplitude for the oscillations of a system under a force which is zero for $t < 0$, $F_0 t/T$ for $0 < t < T$, and F_0 for $t > T$ (Fig. 24), if up to time $t = 0$ the system is at rest in equilibrium.

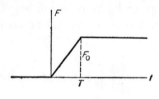

FIG. 24

SOLUTION. During the interval $0 < t < T$ the oscillations are determined by the initial condition as $x = (F_0/mT\omega^3)(\omega t - \sin \omega t)$. For $t > T$ we seek a solution in the form

$$ x = c_1 \cos \omega(t - T) + c_2 \sin \omega(t - T) + F_0/m\omega^2. $$

The continuity of x and $\dot{x}$ at $t = T$ gives $c_1 = -(F_0/mT\omega^3) \sin \omega T$, $c_2 = (F_0/mT\omega^3) \times (1 - \cos \omega T)$. The amplitude is $a = \sqrt{(c_1^2 + c_2^2)} = (2F_0/mT\omega^3) \sin \frac{1}{2}\omega T$. This is the smaller, the more slowly the force F_0 is applied (i.e. the greater T).

PROBLEM 3. The same as Problem 2, but for a constant force F_0 which acts for a finite time T (Fig. 25).

SOLUTION. As in Problem 2, or more simply by using formula (22.10). For $t > T$ we have free oscillations about $x = 0$, and

$$\xi = \frac{F_0}{m} \exp(i\omega t) \int_0^T \exp(-i\omega t)\, dt$$

$$= \frac{F_0}{i\omega m}[1 - \exp(-i\omega T)]\exp(i\omega t).$$

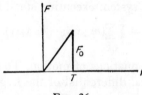

FIG. 25

The squared modulus of ξ gives the amplitude from the relation $|\xi|^2 = a^2\omega^2$. The result is

$$a = (2F_0/m\omega^2)\sin\tfrac{1}{2}\omega T.$$

PROBLEM 4. The same as Problem 2, but for a force $F_0 t/T$ which acts between $t = 0$ and $t = T$ (Fig. 26).

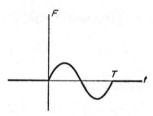

FIG. 26

SOLUTION. By the same method we obtain

$$a = (F_0/Tm\omega^3)\sqrt{[\omega^2 T^2 - 2\omega T\sin\omega T + 2(1 - \cos\omega T)]}.$$

PROBLEM 5. The same as Problem 2, but for a force $F_0\sin\omega t$ which acts between $t = 0$ and $t = T \equiv 2\pi/\omega$ (Fig. 27).

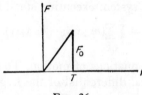

FIG. 27

SOLUTION. Substituting in (22.10) $F(t) = F_0\sin\omega t = F_0[\exp(i\omega t) - \exp(-i\omega t)]/2i$ and integrating from 0 to T, we obtain $a = F_0\pi/m\omega^2$.

§23. Oscillations of systems with more than one degree of freedom

The theory of free oscillations of systems with s degrees of freedom is analogous to that given in §21 for the case $s = 1$.

Let the potential energy of the system U as a function of the generalised co-ordinates q_i $(i = 1, 2, ..., s)$ have a minimum for $q_i = q_{i0}$. Putting

$$x_i = q_i - q_{i0} \tag{23.1}$$

for the small displacements from equilibrium and expanding U as a function of the x_i as far as the quadratic terms, we obtain the potential energy as a positive definite quadratic form

$$U = \tfrac{1}{2} \sum_{i,k} k_{ik} x_i x_k, \tag{23.2}$$

where we again take the minimum value of the potential energy as zero. Since the coefficients k_{ik} and k_{ki} in (23.2) multiply the same quantity $x_i x_k$, it is clear that they may always be considered equal: $k_{ik} = k_{ki}$.

In the kinetic energy, which has the general form $\tfrac{1}{2}\sum a_{ik}(q)\dot{q}_i \dot{q}_k$ (see (5.5)), we put $q_i = q_{i0}$ in the coefficients a_{ik} and, denoting $a_{ik}(q_0)$ by m_{ik}, obtain the kinetic energy as a positive definite quadratic form

$$\tfrac{1}{2} \sum_{i,k} m_{ik} \dot{x}_i \dot{x}_k. \tag{23.3}$$

The coefficients m_{ik} also may always be regarded as symmetrical: $m_{ik} = m_{ki}$. Thus the Lagrangian of a system executing small free oscillations is

$$L = \tfrac{1}{2} \sum_{i,k} (m_{ik} \dot{x}_i \dot{x}_k - k_{ik} x_i x_k). \tag{23.4}$$

Let us now derive the equations of motion. To determine the derivatives involved, we write the total differential of the Lagrangian:

$$dL = \tfrac{1}{2} \sum_{i,k} (m_{ik} \dot{x}_i \, d\dot{x}_k + m_{ik} \dot{x}_k \, d\dot{x}_i - k_{ik} x_i \, dx_k - k_{ik} x_k dx_i).$$

Since the value of the sum is obviously independent of the naming of the suffixes, we can interchange i and k in the first and third terms in the parentheses. Using the symmetry of m_{ik} and k_{ik}, we have

$$dL = \sum (m_{ik} \dot{x}_k d\dot{x}_i - k_{ik} x_k \, dx_i).$$

Hence

$$\partial L/\partial \dot{x}_i = \sum_k m_{ik} \dot{x}_k, \qquad \partial L/\partial x_i = - \sum_k k_{ik} x_k.$$

Lagrange's equations are therefore

$$\sum_k m_{ik} \ddot{x}_k + \sum_k k_{ik} x_k = 0 \qquad (i = 1, 2, ..., s); \tag{23.5}$$

they form a set of s linear homogeneous differential equations with constant coefficients.

As usual, we seek the s unknown functions $x_k(t)$ in the form

$$x_k = A_k \exp(i\omega t), \tag{23.6}$$

where A_k are some constants to be determined. Substituting (23.6) in the

equations (23.5) and cancelling $\exp(i\omega t)$, we obtain a set of linear homogeneous algebraic equations to be satisfied by the A_k:

$$\sum_k (-\omega^2 m_{ik} + k_{ik})A_k = 0. \tag{23.7}$$

If this system has non-zero solutions, the determinant of the coefficients
must vanish:

$$|k_{ik} - \omega^2 m_{ik}| = 0. \tag{23.8}$$

This is the *characteristic equation* and is of degree s in ω^2. In general, it has
s different real positive roots ω_α^2 ($\alpha = 1, 2, ..., s$); in particular cases, some of
these roots may coincide. The quantities ω_α thus determined are the *characteristic frequencies* or *eigenfrequencies* of the system.

It is evident from physical arguments that the roots of equation (23.8) are
real and positive. For the existence of an imaginary part of ω would mean
the presence, in the time dependence of the co-ordinates x_k (23.6), and so
of the velocities $\dot{x}_k$, of an exponentially decreasing or increasing factor. Such
a factor is inadmissible, since it would lead to a time variation of the total
energy $E = U + T$ of the system, which would therefore not be conserved.

The same result may also be derived mathematically. Multiplying equation
(23.7) by A_i^* and summing over i, we have $\sum(-\omega^2 m_{ik} + k_{ik})A_i^* A_k = 0$,
whence $\omega^2 = \sum k_{ik}A_i^* A_k / \sum m_{ik}A_i^* A_k$. The quadratic forms in the numerator
and denominator of this expression are real, since the coefficients k_{ik} and
m_{ik} are real and symmetrical: $(\sum k_{ik}A_i^* A_k)^* = \sum k_{ik}A_i A_k^* = \sum k_{ki}A_i A_k^*$
$= \sum k_{ik}A_k A_i^*$. They are also positive, and therefore ω^2 is positive.†

The frequencies ω_α having been found, we substitute each of them in
equations (23.7) and find the corresponding coefficients A_k. If all the roots
ω_α of the characteristic equation are different, the coefficients A_k are proportional to the minors of the determinant (23.8) with $\omega = \omega_\alpha$. Let these
minors be $\Delta_{k\alpha}$. A particular solution of the differential equations (23.5) is
therefore $x_k = \Delta_{k\alpha}C_\alpha \exp(i\omega_\alpha t)$, where C_α is an arbitrary complex constant.

The general solution is the sum of s particular solutions. Taking the real
part, we write

$$x_k = \text{re} \sum_{\alpha=1}^{s} \Delta_{k\alpha}C_\alpha \exp(i\omega_\alpha t) \equiv \sum_\alpha \Delta_{k\alpha}\Theta_\alpha, \tag{23.9}$$

where

$$\Theta_\alpha = \text{re}[C_\alpha \exp(i\omega_\alpha t)]. \tag{23.10}$$

Thus the time variation of each co-ordinate of the system is a super-
position of s simple periodic oscillations $\Theta_1, \Theta_2, ..., \Theta_s$ with arbitrary amplitudes and phases but definite frequencies.

---

† The fact that a quadratic form with the coefficients k_{ik} is positive definite is seen from
their definition (23.2) for real values of the variables. If the complex quantities A_k are written
explicitly as $a_k + ib_k$, we have, again using the symmetry of k_{ik}, $\sum k_{ik}A_i^* A_k = \sum k_{ik}(a_i - ib_i) \times$
$\times (a_k + ib_k) = \sum k_{ik}a_i a_k + \sum k_{ik}b_i b_k$, which is the sum of two positive definite forms.

The question naturally arises whether the generalised co-ordinates can be chosen in such a way that each of them executes only one simple oscillation. The form of the general integral (23.9) points to the answer. For, regarding the s equations (23.9) as a set of equations for s unknowns Θ_α, we can express $\Theta_1, \Theta_2, ..., \Theta_s$ in terms of the co-ordinates $x_1, x_2, ..., x_s$. The quantities Θ_α may therefore be regarded as new generalised co-ordinates, called *normal co-ordinates*, and they execute simple periodic oscillations, called *normal oscillations* of the system.

The normal co-ordinates Θ_α are seen from their definition to satisfy the equations

$$\ddot{\Theta}_\alpha + \omega_\alpha^2 \Theta_\alpha = 0. \tag{23.11}$$

This means that in normal co-ordinates the equations of motion become s independent equations. The acceleration in each normal co-ordinate depends only on the value of that co-ordinate, and its time dependence is entirely determined by the initial values of the co-ordinate and of the corresponding velocity. In other words, the normal oscillations of the system are completely independent.

It is evident that the Lagrangian expressed in terms of normal co-ordinates is a sum of expressions each of which corresponds to oscillation in one dimension with one of the frequencies ω_α, i.e. it is of the form

$$L = \sum_\alpha \tfrac{1}{2} m_\alpha (\dot{\Theta}_\alpha^2 - \omega_\alpha^2 \Theta_\alpha^2), \tag{23.12}$$

where the m_α are positive constants. Mathematically, this means that the transformation (23.9) simultaneously puts both quadratic forms—the kinetic energy (23.3) and the potential energy (23.2)—in diagonal form.

The normal co-ordinates are usually chosen so as to make the coefficients of the squared velocities in the Lagrangian equal to one-half. This can be achieved by simply defining new normal co-ordinates Q_α by

$$Q_\alpha = \sqrt{m_\alpha} \Theta_\alpha. \tag{23.13}$$

Then

$$L = \tfrac{1}{2} \sum_\alpha (\dot{Q}_\alpha^2 - \omega_\alpha^2 Q_\alpha^2).$$

The above discussion needs little alteration when some roots of the characteristic equation coincide. The general form (23.9), (23.10) of the integral of the equations of motion remains unchanged, with the same number s of terms, and the only difference is that the coefficients $\Delta_{k\alpha}$ corresponding to multiple roots are not the minors of the determinant, which in this case vanish.†

---

† The impossibility of terms in the general integral which contain powers of the time as well as the exponential factors is seen from the same argument as that which shows that the frequencies are real: such terms would violate the law of conservation of energy.

Each multiple (or, as we say, *degenerate*) frequency corresponds to a number of normal co-ordinates equal to its multiplicity, but the choice of these co-ordinates is not unique. The normal co-ordinates with equal ω_α enter the kinetic and potential energies as sums $\Sigma \dot{Q}_\alpha{}^2$ and $\Sigma Q_\alpha{}^2$ which are transformed in the same way, and they can be linearly transformed in any manner which does not alter these sums of squares.

The normal co-ordinates are very easily found for three-dimensional oscillations of a single particle in a constant external field. Taking the origin of Cartesian co-ordinates at the point where the potential energy $U(x, y, z)$ is a minimum, we obtain this energy as a quadratic form in the variables x, y, z, and the kinetic energy $T = \frac{1}{2}m(\dot{x}^2+\dot{y}^2+\dot{z}^2)$ (where m is the mass of the particle) does not depend on the orientation of the co-ordinate axes. We therefore have only to reduce the potential energy to diagonal form by an appropriate choice of axes. Then

$$L = \tfrac{1}{2}m(\dot{x}^2+\dot{y}^2+\dot{z}^2)-\tfrac{1}{2}(k_1x^2+k_2y^2+k_3z^2), \tag{23.14}$$

and the normal oscillations take place in the x, y and z directions with frequencies $\omega_1 = \sqrt{(k_1/m)}$, $\omega_2 = \sqrt{(k_2/m)}$, $\omega_3 = \sqrt{(k_3/m)}$. In the particular case of a central field ($k_1 = k_2 = k_3 \equiv k$, $U = \frac{1}{2}kr^2$) the three frequencies are equal (see Problem 3).

The use of normal co-ordinates makes possible the reduction of a problem of forced oscillations of a system with more than one degree of freedom to a series of problems of forced oscillation in one dimension. The Lagrangian of the system, including the variable external forces, is

$$L = L_0 + \sum_k F_k(t)x_k, \tag{23.15}$$

where L_0 is the Lagrangian for free oscillations. Replacing the co-ordinates x_k by normal co-ordinates, we have

$$L = \tfrac{1}{2}\sum_\alpha(\dot{Q}_\alpha{}^2 - \omega_\alpha{}^2 Q_\alpha{}^2)+ \sum_\alpha f_\alpha(t)Q_\alpha, \tag{23.16}$$

where we have put

$$f_\alpha(t) = \sum_k F_k(t)\Delta_{k\alpha}/\sqrt{m_\alpha}.$$

The corresponding equations of motion

$$\ddot{Q}_\alpha+ \omega_\alpha{}^2 Q_\alpha = f_\alpha(t) \tag{23.17}$$

each involve only one unknown function $Q_\alpha(t)$.

PROBLEMS

PROBLEM 1. Determine the oscillations of a system with two degrees of freedom whose Lagrangian is $L = \frac{1}{2}(\dot{x}^2+\dot{y}^2)-\frac{1}{2}\omega_0{}^2(x^2+y^2)+\alpha xy$ (two identical one-dimensional systems of eigenfrequency ω_0 coupled by an interaction $-\alpha xy$).

SOLUTION. The equations of motion are $\ddot{x} + \omega_0^2 x = \alpha y$, $\ddot{y} + \omega_0^2 y = \alpha x$. The substitution (23.6) gives

$$A_x(\omega_0^2 - \omega^2) = \alpha A_y, \qquad A_y(\omega_0^2 - \omega^2) = \alpha A_x. \tag{1}$$

The characteristic equation is $(\omega_0^2 - \omega^2)^2 = \alpha^2$, whence $\omega_1^2 = \omega_0^2 - \alpha$, $\omega_2^2 = \omega_0^2 + \alpha$. For $\omega = \omega_1$, the equations (1) give $A_x = A_y$, and for $\omega = \omega_2$, $A_x = -A_y$. Hence $x = (Q_1 + Q_2)/\sqrt{2}$, $y = (Q_1 - Q_2)/\sqrt{2}$, the coefficients $1/\sqrt{2}$ resulting from the normalisation of the normal co-ordinates as in equation (23.13).

For $\alpha \ll \omega_0^2$ (weak coupling) we have $\omega_1 \simeq \omega_0 \left| - \tfrac{1}{2}\alpha/\omega_0 \right.$, $\omega_2 \simeq \omega_0 \left| + \tfrac{1}{2}\alpha/\omega_0 \right.$. The variation of x and y is in this case a superposition of two oscillations with almost equal frequencies, i.e. beats of frequency $\omega_2 - \omega_1 = \alpha/\omega_0$ (see §22). The amplitude of y is a minimum when that of x is a maximum, and *vice versa*.

PROBLEM 2. Determine the small oscillations of a coplanar double pendulum (Fig. 1, §5).

SOLUTION. For small oscillations ($\phi_1 \ll 1$, $\phi_2 \ll 1$), the Lagrangian derived in §5, Problem 1, becomes

$$L = \tfrac{1}{2}(m_1 + m_2)l_1^2\dot{\phi}_1^2 + \tfrac{1}{2}m_2 l_2^2\dot{\phi}_2^2 + m_2 l_1 l_2\dot{\phi}_1\dot{\phi}_2 - \tfrac{1}{2}(m_1 + m_2)gl_1\phi_1^2 - \tfrac{1}{2}m_2 g l_2\phi_2^2.$$

The equations of motion are

$$(m_1 + m_2)l_1\ddot{\phi}_1 + m_2 l_2\ddot{\phi}_2 + (m_1 + m_2)g\phi_1 = 0, \qquad l_1\ddot{\phi}_1 + l_2\ddot{\phi}_2 + g\phi_2 = 0.$$

Substitution of (23.6) gives

$$A_1(m_1 + m_2)(g - l_1\omega^2) - A_2\omega^2 m_2 l_2 = 0, \qquad -A_1 l_1\omega^2 + A_2(g - l_2\omega^2) = 0.$$

The roots of the characteristic equation are

$$\omega_{1,2}^2 = \frac{g}{2m_1 l_1 l_2}\{(m_1 + m_2)(l_1 + l_2) \pm \sqrt{(m_1 + m_2)}\sqrt{[(m_1 + m_2)(l_1 + l_2)^2 - 4m_1 l_1 l_2]}\}.$$

As $m_1 \to \infty$ the frequencies tend to the values $\sqrt{(g/l_1)}$ and $\sqrt{(g/l_2)}$, corresponding to independent oscillations of the two pendulums.

PROBLEM 3. Find the path of a particle in a central field $U = \tfrac{1}{2}kr^2$ (called a *space oscillator*).

SOLUTION. As in any central field, the path lies in a plane, which we take as the xy-plane. The variation of each co-ordinate x, y is a simple oscillation with the same frequency $\omega = \sqrt{(k/m)}$: $x = a\cos(\omega t + \alpha)$, $y = b\cos(\omega t + \beta)$, or $x = a\cos\phi$, $y = b\cos(\phi + \delta) = b\cos\delta\cos\phi - b\sin\delta\sin\phi$, where $\phi = \omega t + \alpha$, $\delta = \beta - \alpha$. Solving for $\cos\phi$ and $\sin\phi$ and equating the sum of their squares to unity, we find the equation of the path:

$$\frac{x^2}{a^2} + \frac{y^2}{b^2} - \frac{2xy}{ab}\cos\delta = \sin^2\delta.$$

This is an ellipse with its centre at the origin.† When $\delta = 0$ or π, the path degenerates to a segment of a straight line.

§24. Vibrations of molecules

If we have a system of interacting particles not in an external field, not all of its degrees of freedom relate to oscillations. A typical example is that of molecules. Besides motions in which the atoms oscillate about their positions of equilibrium in the molecule, the whole molecule can execute translational and rotational motions.

Three degrees of freedom correspond to translational motion, and in general the same number to rotation, so that, of the $3n$ degrees of freedom of a molecule containing n atoms, $3n - 6$ correspond to vibration. An exception is formed

---

† The fact that the path in a field with potential energy $U = \tfrac{1}{2}kr^2$ is a closed curve has already been mentioned in §14.

by molecules in which the atoms are collinear, for which there are only two rotational degrees of freedom (since rotation about the line of atoms is of no significance), and therefore $3n-5$ vibrational degrees of freedom.

In solving a mechanical problem of molecular oscillations, it is convenient to eliminate immediately the translational and rotational degrees of freedom. The former can be removed by equating to zero the total momentum of the molecule. Since this condition implies that the centre of mass of the molecule is at rest, it can be expressed by saying that the three co-ordinates of the centre of mass are constant. Putting $\mathbf{r}_a = \mathbf{r}_{a0} + \mathbf{u}_a$, where $\mathbf{r}_{a0}$ is the radius vector of the equilibrium position of the ath atom, and $\mathbf{u}_a$ its deviation from this position, we have the condition $\sum m_a \mathbf{r}_a = \text{constant} = \sum m_a \mathbf{r}_{a0}$ or

$$\sum m_a \mathbf{u}_a = 0. \tag{24.1}$$

To eliminate the rotation of the molecule, its total angular momentum must be equated to zero. Since the angular momentum is not the total time derivative of a function of the co-ordinates, the condition that it is zero cannot in general be expressed by saying that some such function is zero. For small oscillations, however, this can in fact be done. Putting again $\mathbf{r}_a = \mathbf{r}_{a0} + \mathbf{u}_a$ and neglecting small quantities of the second order in the displacements $\mathbf{u}_a$, we can write the angular momentum of the molecule as

$$\mathbf{M} = \sum m_a \mathbf{r}_a \times \mathbf{v}_a \cong \sum m_a \mathbf{r}_{a0} \times \dot{\mathbf{u}}_a = (d/dt) \sum m_a \mathbf{r}_{a0} \times \mathbf{u}_a.$$

The condition for this to be zero is therefore, in the same approximation,

$$\sum m_a \mathbf{r}_{a0} \times \mathbf{u}_a = 0, \tag{24.2}$$

in which the origin may be chosen arbitrarily.

The normal vibrations of the molecule may be classified according to the corresponding motion of the atoms on the basis of a consideration of the symmetry of the equilibrium positions of the atoms in the molecule. There is a general method of doing so, based on the use of group theory, which we discuss elsewhere.† Here we shall consider only some elementary examples.

If all n atoms in a molecule lie in one plane, we can distinguish normal vibrations in which the atoms remain in that plane from those where they do not. The number of each kind is readily determined. Since, for motion in a plane, there are $2n$ degrees of freedom, of which two are translational and one rotational, the number of normal vibrations which leave the atoms in the plane is $2n-3$. The remaining $(3n-6)-(2n-3) = n-3$ vibrational degrees of freedom correspond to vibrations in which the atoms move out of the plane.

For a linear molecule we can distinguish longitudinal vibrations, which maintain the linear form, from vibrations which bring the atoms out of line. Since a motion of n particles in a line corresponds to n degrees of freedom, of which one is translational, the number of vibrations which leave the atoms

† See *Quantum Mechanics*, §100, Pergamon Press, Oxford 1976.

in line is $n-1$. Since the total number of vibrational degrees of freedom of a linear molecule is $3n-5$, there are $2n-4$ which bring the atoms out of line. These $2n-4$ vibrations, however, correspond to only $n-2$ different frequencies, since each such vibration can occur in two mutually perpendicular planes through the axis of the molecule. It is evident from symmetry that each such pair of normal vibrations have equal frequencies.

PROBLEMS†

PROBLEM 1. Determine the frequencies of vibrations of a symmetrical linear triatomic molecule ABA (Fig. 28). It is assumed that the potential energy of the molecule depends only on the distances AB and BA and the angle ABA.

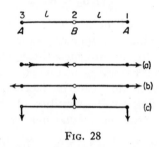

FIG. 28

SOLUTION. The longitudinal displacements x_1, x_2, x_3 of the atoms are related, according to (24.1), by $m_A(x_1+x_3)+m_Bx_2 = 0$. Using this, we eliminate x_2 from the Lagrangian of the longitudinal motion

$$L = \tfrac{1}{2}m_A(\dot{x}_1{}^2+\dot{x}_3{}^2)+\tfrac{1}{2}m_B\dot{x}_2{}^2-\tfrac{1}{2}k_1[(x_1-x_2)^2+(x_3-x_2)^2],$$

and use new co-ordinates $Q_a = x_1+x_3$, $Q_s = x_1-x_3$. The result is

$$L = \frac{\mu m_A}{4m_B}\dot{Q}_a{}^2+\frac{m_A}{4}\dot{Q}_s{}^2-\frac{k_1\mu^2}{4m_B{}^2}Q_a{}^2-\frac{k_1}{4}Q_s{}^2,$$

where $\mu = 2m_A+m_B$ is the mass of the molecule. Hence we see that Q_a and Q_s are normal co-ordinates (not yet normalised). The co-ordinate Q_a corresponds to a vibration anti-symmetrical about the centre of the molecule ($x_1 = x_3$; Fig. 28a), with frequency $\omega_a = \sqrt{(k_1\mu/m_Am_B)}$. The co-ordinate Q_s corresponds to a symmetrical vibration ($x_1 = -x_3$; Fig. 28b), with frequency $\omega_{s1} = \sqrt{(k_1/m_A)}$.

The transverse displacements y_1, y_2, y_3 of the atoms are, according to (24.1) and (24.2), related by $m_A(y_1+y_2)+m_By_2 = 0$, $y_1 = y_3$ (a symmetrical bending of the molecule; Fig. 28c). The potential energy of this vibration can be written as $\tfrac{1}{2}k_2l^2\delta^2$, where δ is the deviation of the angle ABA from the value π, given in terms of the displacements by $\delta = [(y_1-y_2)+(y_3-y_2)]/l$. Expressing y_1, y_2, y_3 in terms of δ, we obtain the Lagrangian of the transverse motion:

$$L = \tfrac{1}{2}m_A(\dot{y}_1{}^2+\dot{y}_3{}^2)+\tfrac{1}{2}m_B\dot{y}_2{}^2-\tfrac{1}{2}k_2l^2\delta^2$$
$$= \frac{m_Am_B}{4\mu}l^2\dot{\delta}^2-\tfrac{1}{2}k_2l^2\delta^2,$$

whence the frequency is $\omega_{s2} = \sqrt{(2k_2\mu/m_Am_B)}$.

---

† Calculations of the vibrations of more complex molecules are given by M. V. VOL'KENSH-TEĬN, M. A. EL'YASHEVICH and B. I. STEPANOV, *Molecular Vibrations (Kolebaniya molekul),* Moscow 1949; G. HERZBERG, *Molecular Spectra and Molecular Structure: Infra-red and Raman Spectra of Polyatomic Molecules,* Van Nostrand, New York 1945.

PROBLEM 2. The same as Problem 1, but for a triangular molecule ABA (Fig. 29).

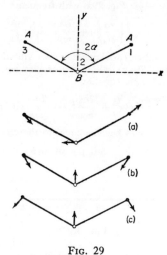

FIG. 29

SOLUTION. By (24.1) and (24.2) the x and y components of the displacements $\mathbf{u}$ of the atoms are related by

$$m_A(x_1+x_3)+m_Bx_2 = 0,$$
$$m_A(y_1+y_3)+m_By_2 = 0,$$
$$(y_1-y_3)\sin\alpha-(x_1+x_3)\cos\alpha = 0.$$

The changes δl_1 and δl_2 in the distances AB and BA are obtained by taking the components along these lines of the vectors $\mathbf{u}_1-\mathbf{u}_2$ and $\mathbf{u}_3-\mathbf{u}_2$:

$$\delta l_1 = (x_1-x_2)\sin\alpha+(y_1-y_2)\cos\alpha,$$
$$\delta l_2 = -(x_3-x_2)\sin\alpha+(y_3-y_2)\cos\alpha.$$

The change in the angle ABA is obtained by taking the components of those vectors perpendicular to AB and BA:

$$\delta = \frac{1}{l}[(x_1-x_2)\cos\alpha-(y_1-y_2)\sin\alpha]+\frac{1}{l}[-(x_3-x_2)\cos\alpha-(y_3-y_2)\sin\alpha].$$

The Lagrangian of the molecule is

$$L = \tfrac{1}{2}m_A(\dot{\mathbf{u}}_1{}^2+\dot{\mathbf{u}}_3{}^2)+\tfrac{1}{2}m_B\dot{\mathbf{u}}_2{}^2-\tfrac{1}{2}k_1(\delta l_1{}^2+\delta l_2{}^2)-\tfrac{1}{2}k_2l^2\delta^2.$$

We use the new co-ordinates $Q_a = x_1+x_3$, $q_{s1} = x_1-x_3$, $q_{s2} = y_1+y_3$. The components of the vectors $\mathbf{u}$ are given in terms of these co-ordinates by $x_1 = \tfrac{1}{2}(Q_a+q_{s1})$, $x_3 = \tfrac{1}{2}(Q_a-q_{s1})$, $x_2 = -m_AQ_a/m_B$, $y_1 = \tfrac{1}{2}(q_{s2}+Q_a\cot\alpha)$, $y_3 = \tfrac{1}{2}(q_{s2}-Q_a\cot\alpha)$, $y_2 = -m_Aq_{s2}/m_B$. The Lagrangian becomes

$$L = \tfrac{1}{4}m_A\left(\frac{2m_A}{m_B}+\frac{1}{\sin^2\alpha}\right)\dot{Q}_a{}^2+\tfrac{1}{4}m_A\dot{q}_{s1}{}^2+\frac{\mu m_A}{4m_B}\dot{q}_{s2}{}^2-$$

$$-\tfrac{1}{4}k_1Q_a{}^2\left(\frac{2m_A}{m_B}+\frac{1}{\sin^2\alpha}\right)\left(1+\frac{2m_A}{m_B}\sin^2\alpha\right)-$$

$$-\tfrac{1}{4}q_{s1}{}^2(k_1\sin^2\alpha+2k_2\cos^2\alpha)-\tfrac{1}{4}q_{s2}{}^2\frac{\mu^2}{m_B{}^2}(k_1\cos^2\alpha+2k_2\sin^2\alpha)+$$

$$+q_{s1}q_{s2}\frac{\mu}{2m_B}(2k_2-k_1)\sin\alpha\cos\alpha.$$

Hence we see that the co-ordinate Q_a corresponds to a normal vibration antisymmetrical about the y-axis ($x_1 = x_3$, $y_1 = -y_3$; Fig. 29a) with frequency

$$\omega_a = \sqrt{\left[\frac{k_1}{m_A}\left(1 + \frac{2m_A}{m_B}\sin^2\alpha\right)\right]}.$$

The co-ordinates q_{s1}, q_{s2} together correspond to two vibrations symmetrical about the y-axis ($x_1 = -x_3$, $y_1 = y_3$; Fig. 29b, c), whose frequencies ω_{s1}, ω_{s2} are given by the roots of the quadratic (in ω^2) characteristic equation

$$\omega^4 - \omega^2\left[\frac{k_1}{m_A}\left(1 + \frac{2m_A}{m_B}\cos^2\alpha\right) + \frac{2k_2}{m_A}\left(1 + \frac{2m_A}{m_B}\sin^2\alpha\right)\right] + \frac{2\mu k_1 k_2}{m_A{}^2 m_B} = 0.$$

When $2\alpha = \pi$, all three frequencies become equal to those derived in Problem 1.

PROBLEM 3. The same as Problem 1, but for an unsymmetrical linear molecule ABC (Fig. 30).

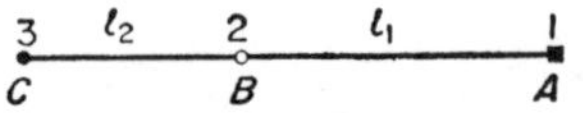

FIG. 30

SOLUTION. The longitudinal (x) and transverse (y) displacements of the atoms are related by

$$m_A x_1 + m_B x_2 + m_C x_3 = 0, \qquad m_A y_1 + m_B y_2 + m_C y_3 = 0,$$
$$m_A l_1 y_1 = m_C l_2 y_3.$$

The potential energy of stretching and bending can be written as $\frac{1}{2}k_1(\delta l_1)^2 + \frac{1}{2}k_1'(\delta l_2)^2 + \frac{1}{2}k_2 l^2\delta^2$, where $2l = l_1 + l_2$. Calculations similar to those in Problem 1 give

$$\omega_t{}^2 = \frac{k_2 l^2}{l_1{}^2 l_2{}^2}\left(\frac{l_1{}^2}{m_C} + \frac{l_2{}^2}{m_A} + \frac{4l^2}{m_B}\right)$$

for the transverse vibrations and the quadratic (in ω^2) equation

$$\omega^4 - \omega^2\left[k_1\left(\frac{1}{m_A} + \frac{1}{m_B}\right) + k_1'\left(\frac{1}{m_B} + \frac{1}{m_C}\right)\right] + \frac{\mu k_1 k_1'}{m_A m_B m_C} = 0$$

for the frequencies ω_{l1}, ω_{l2} of the longitudinal vibrations.

§25. Damped oscillations

So far we have implied that all motion takes place in a vacuum, or else that the effect of the surrounding medium on the motion may be neglected. In reality, when a body moves in a medium, the latter exerts a resistance which tends to retard the motion. The energy of the moving body is finally dissipated by being converted into heat.

Motion under these conditions is no longer a purely mechanical process, and allowance must be made for the motion of the medium itself and for the internal thermal state of both the medium and the body. In particular, we cannot in general assert that the acceleration of a moving body is a function only of its co-ordinates and velocity at the instant considered; that is, there are no equations of motion in the mechanical sense. Thus the problem of the motion of a body in a medium is not one of mechanics.

There exists, however, a class of cases where motion in a medium can be approximately described by including certain additional terms in the mechanical equations of motion. Such cases include oscillations with frequencies small compared with those of the dissipative processes in the medium. When this condition is fulfilled we may regard the body as being acted on by a force of *friction* which depends (for a given homogeneous medium) only on its velocity.

If, in addition, this velocity is sufficiently small, then the frictional force can be expanded in powers of the velocity. The zero-order term in the expansion is zero, since no friction acts on a body at rest, and so the first non-vanishing term is proportional to the velocity. Thus the generalised frictional force f_{fr} acting on a system executing small oscillations in one dimension (co-ordinate x) may be written $f_{\mathrm{fr}} = -\alpha\dot{x}$, where α is a positive coefficient and the minus sign indicates that the force acts in the direction opposite to that of the velocity. Adding this force on the right-hand side of the equation of motion, we obtain (see (21.4))

$$m\ddot{x} = -kx - \alpha\dot{x}. \tag{25.1}$$

We divide this by m and put

$$k/m = \omega_0{}^2, \qquad \alpha/m = 2\lambda; \tag{25.2}$$

ω_0 is the frequency of free oscillations of the system in the absence of friction, and λ is called the *damping coefficient* or *damping decrement*.†

Thus the equation is

$$\ddot{x} + 2\lambda\dot{x} + \omega_0{}^2 x = 0. \tag{25.3}$$

We again seek a solution $x = \exp(rt)$ and obtain r for the characteristic equation $r^2 + 2\lambda r + \omega_0{}^2 = 0$, whence $r_{1,2} = -\lambda \pm \sqrt{(\lambda^2 - \omega_0{}^2)}$. The general solution of equation (25.3) is

$$x = c_1 \exp(r_1 t) + c_2 \exp(r_2 t).$$

Two cases must be distinguished. If $\lambda < \omega_0$, we have two complex conjugate values of r. The general solution of the equation of motion can then be written as

$$x = \mathrm{re}\{A \exp[-\lambda t + i\sqrt{(\omega_0{}^2 - \lambda^2)}t]\},$$

where A is an arbitrary complex constant, or as

$$x = a \exp(-\lambda t) \cos(\omega t + \alpha), \tag{25.4}$$

with $\omega = \sqrt{(\omega_0{}^2 - \lambda^2)}$ and a and α real constants. The motion described by these formulae consists of *damped oscillations*. It may be regarded as being harmonic oscillations of exponentially decreasing amplitude. The rate of decrease of the amplitude is given by the exponent λ, and the "frequency" ω is less than that of free oscillations in the absence of friction. For $\lambda \ll \omega_0$, the difference between ω and ω_0 is of the second order of smallness. The decrease in frequency as a result of friction is to be expected, since friction retards motion.

---

† The dimensionless product λT (where $T = 2\pi/\omega$ is the period) is called the *logarithmic damping decrement*.

If $\lambda \ll \omega_0$, the amplitude of the damped oscillation is almost unchanged during the period $2\pi/\omega$. It is then meaningful to consider the mean values (over the period) of the squared co-ordinates and velocities, neglecting the change in $\exp(-\lambda t)$ when taking the mean. These mean squares are evidently proportional to $\exp(-2\lambda t)$. Hence the mean energy of the system decreases as

$$\bar{E} = E_0 \exp(-2\lambda t), \tag{25.5}$$

where E_0 is the initial value of the energy.

Next, let $\lambda > \omega_0$. Then the values of r are both real and negative. The general form of the solution is

$$x = c_1 \exp\{-[\lambda - \sqrt{(\lambda^2 - \omega_0^2)}]t\} + c_2 \exp\{-[\lambda + \sqrt{(\lambda^2 - \omega_0^2)}]t\}. \tag{25.6}$$

We see that in this case, which occurs when the friction is sufficiently strong, the motion consists of a decrease in $|x|$, i.e. an asymptotic approach (as $t \to \infty$) to the equilibrium position. This type of motion is called *aperiodic damping*.

Finally, in the special case where $\lambda = \omega_0$, the characteristic equation has the double root $r = -\lambda$. The general solution of the differential equation is then

$$x = (c_1 + c_2 t) \exp(-\lambda t). \tag{25.7}$$

This is a special case of aperiodic damping.

For a system with more than one degree of freedom, the generalised frictional forces corresponding to the co-ordinates x_i are linear functions of the velocities, of the form

$$f_{\mathrm{fr},i} = -\sum_k \alpha_{ik} \dot{x}_k. \tag{25.8}$$

From purely mechanical arguments we can draw no conclusions concerning the symmetry properties of the coefficients α_{ik} as regards the suffixes i and k, but the methods of statistical physics† make it possible to demonstrate that in all cases

$$\alpha_{ik} = \alpha_{ki}. \tag{25.9}$$

Hence the expressions (25.8) can be written as the derivatives

$$f_{\mathrm{fr},i} = -\partial F/\partial \dot{x}_i \tag{25.10}$$

of the quadratic form

$$F = \tfrac{1}{2} \sum_{i,k} \alpha_{ik} \dot{x}_i \dot{x}_k, \tag{25.11}$$

which is called the *dissipative function*.

The forces (25.10) must be added to the right-hand side of Lagrange's equations:

$$\frac{\mathrm{d}}{\mathrm{d}t}\left(\frac{\partial L}{\partial \dot{x}_i}\right) = \frac{\partial L}{\partial x_i} - \frac{\partial F}{\partial \dot{x}_i}. \tag{25.12}$$

---

† See *Statistical Physics*, §123, Pergamon Press, Oxford 1969.

The dissipative function itself has an important physical significance: it gives the rate of dissipation of energy in the system. This is easily seen by calculating the time derivative of the mechanical energy of the system. We have

$$\frac{dE}{dt} = \frac{d}{dt}\left(\sum_i \dot{x}_i \frac{\partial L}{\partial \dot{x}_i} - L\right)$$

$$= \sum_i \dot{x}_i \left(\frac{d}{dt}\left[\frac{\partial L}{\partial \dot{x}_i}\right] - \frac{\partial L}{\partial x_i}\right)$$

$$= -\sum_i \dot{x}_i \frac{\partial F}{\partial \dot{x}_i}.$$

Since F is a quadratic function of the velocities, Euler's theorem on homogeneous functions shows that the sum on the right-hand side is equal to $2F$. Thus

$$dE/dt = -2F, \tag{25.13}$$

i.e. the rate of change of the energy of the system is twice the dissipative function. Since dissipative processes lead to loss of energy, it follows that $F > 0$, i.e. the quadratic form (25.11) is positive definite.

The equations of small oscillations under friction are obtained by adding the forces (25.8) to the right-hand sides of equations (23.5):

$$\sum_k m_{ik}\ddot{x}_k + \sum_k k_{ik}x_k = -\sum_k \alpha_{ik}\dot{x}_k. \tag{25.14}$$

Putting in these equations $x_k = A_k \exp(rt)$, we obtain, on cancelling $\exp(rt)$, a set of linear algebraic equations for the constants A_k:

$$\sum_k (m_{ik}r^2 + \alpha_{ik}r + k_{ik})A_k = 0. \tag{25.15}$$

Equating to zero their determinant, we find the characteristic equation, which determines the possible values of r:

$$|m_{ik}r^2 + \alpha_{ik}r + k_{ik}| = 0. \tag{25.16}$$

This is an equation in r of degree $2s$. Since all the coefficients are real, its roots are either real, or complex conjugate pairs. The real roots must be negative, and the complex roots must have negative real parts, since otherwise the co-ordinates, velocities and energy of the system would increase exponentially with time, whereas dissipative forces must lead to a decrease of the energy.

§26. Forced oscillations under friction

The theory of forced oscillations under friction is entirely analogous to that given in §22 for oscillations without friction. Here we shall consider in detail the case of a periodic external force, which is of considerable interest.

Adding to the right-hand side of equation (25.1) an external force $f \cos \gamma t$ and dividing by m, we obtain the equation of motion:

$$\ddot{x} + 2\lambda\dot{x} + \omega_0^2 x = (f/m) \cos \gamma t. \tag{26.1}$$

The solution of this equation is more conveniently found in complex form, and so we replace $\cos \gamma t$ on the right by $\exp(i\gamma t)$:

$$\ddot{x} + 2\lambda\dot{x} + \omega_0^2 x = (f/m) \exp(i\gamma t).$$

We seek a particular integral in the form $x = B \exp(i\gamma t)$, obtaining for B the value

$$B = f/m(\omega_0^2 - \gamma^2 + 2i\lambda\gamma). \tag{26.2}$$

Writing $B = b \exp(i\delta)$, we have

$$b = f/m\sqrt{[(\omega_0^2 - \gamma^2)^2 + 4\lambda^2\gamma^2]}, \quad \tan \delta = 2\lambda\gamma/(\gamma^2 - \omega_0^2). \tag{26.3}$$

Finally, taking the real part of the expression $B \exp(i\gamma t) = b \exp[i(\gamma t + \delta)]$, we find the particular integral of equation (26.1); adding to this the general solution of that equation with zero on the right-hand side (and taking for definiteness the case $\omega_0 > \lambda$), we have

$$x = a \exp(-\lambda t) \cos(\omega t + \alpha) + b \cos(\gamma t + \delta). \tag{26.4}$$

The first term decreases exponentially with time, so that, after a sufficient time, only the second term remains:

$$x = b \cos(\gamma t + \delta). \tag{26.5}$$

The expression (26.3) for the amplitude b of the forced oscillation increases as γ approaches ω_0, but does not become infinite as it does in resonance without friction. For a given amplitude f of the force, the amplitude of the oscillations is greatest when $\gamma = \sqrt{(\omega_0^2 - 2\lambda^2)}$; for $\lambda \ll \omega_0$, this differs from ω_0 only by a quantity of the second order of smallness.

Let us consider the range near resonance, putting $\gamma = \omega_0 + \epsilon$ with ϵ small, and suppose also that $\lambda \ll \omega_0$. Then we can approximately put, in (26.2), $\gamma^2 - \omega_0^2 = (\gamma + \omega_0)(\gamma - \omega_0) \approx 2\omega_0\epsilon$, $2i\lambda\gamma \approx 2i\lambda\omega_0$, so that

$$B = -f/2m(\epsilon - i\lambda)\omega_0 \tag{26.6}$$

or

$$b = f/2m\omega_0\sqrt{(\epsilon^2 + \lambda^2)}, \quad \tan \delta = \lambda/\epsilon. \tag{26.7}$$

A property of the phase difference δ between the oscillation and the external force is that it is always negative, i.e. the oscillation "lags behind" the force. Far from resonance on the side $\gamma < \omega_0$, $\delta \to 0$; on the side $\gamma > \omega_0$, $\delta \to -\pi$. The change of δ from zero to $-\pi$ takes place in a frequency range near ω_0 which is narrow (of the order of λ in width); δ passes through $-\frac{1}{2}\pi$ when $\gamma = \omega_0$. In the absence of friction, the phase of the forced oscillation changes discontinuously by π at $\gamma = \omega_0$ (the second term in (22.4) changes sign); when friction is allowed for, this discontinuity is smoothed out.

In steady motion, when the system executes the forced oscillations given by (26.5), its energy remains unchanged. Energy is continually absorbed by the system from the source of the external force and dissipated by friction. Let $I(\gamma)$ be the mean amount of energy absorbed per unit time, which depends on the frequency of the external force. By (25.13) we have $I(\gamma) = 2\bar{F}$, where $\bar{F}$ is the average value (over the period of oscillation) of the dissipative function. For motion in one dimension, the expression (25.11) for the dissipative function becomes $F = \frac{1}{2}\alpha \dot{x}^2 = \lambda m \dot{x}^2$. Substituting (26.5), we have

$$F = \lambda m b^2 \gamma^2 \sin^2(\gamma t + \delta).$$

The time average of the squared sine is $\frac{1}{2}$, so that

$$I(\gamma) = \lambda m b^2 \gamma^2. \qquad (26.8)$$

Near resonance we have, on substituting the amplitude of the oscillation from (26.7),

$$I(\epsilon) = f^2 \lambda / 4m(\epsilon^2 + \lambda^2). \qquad (26.9)$$

This is called a *dispersion-type* frequency dependence of the absorption. The *half-width* of the resonance curve (Fig. 31) is the value of $|\epsilon|$ for which $I(\epsilon)$ is half its maximum value ($\epsilon = 0$). It is evident from (26.9) that in the present case the half-width is just the damping coefficient λ. The height of the maximum is $I(0) = f^2/4m\lambda$, and is inversely proportional to λ. Thus,

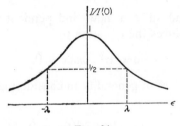

Fig. 31

when the damping coefficient decreases, the resonance curve becomes more peaked. The area under the curve, however, remains unchanged. This area is given by the integral

$$\int_0^\infty I(\gamma) \, \mathrm{d}\gamma = \int_{-\omega_0}^\infty I(\epsilon) \, \mathrm{d}\epsilon.$$

Since $I(\epsilon)$ diminishes rapidly with increasing $|\epsilon|$, the region where $|\epsilon|$ is large is of no importance, and the lower limit may be replaced by $-\infty$, and $I(\epsilon)$ taken to have the form given by (26.9). Then we have

$$\int_{-\infty}^\infty I(\epsilon) \, \mathrm{d}\epsilon = \frac{f^2 \lambda}{4m} \int_{-\infty}^\infty \frac{\mathrm{d}\epsilon}{\epsilon^2 + \lambda^2} = \frac{\pi f^2}{4m}. \qquad (26.10)$$

PROBLEM

Determine the forced oscillations due to an external force $f = f_0 \exp(\alpha t) \cos \gamma t$ in the presence of friction.

SOLUTION. We solve the complex equation of motion

$$\ddot{x} + 2\lambda\dot{x} + \omega_0^2 x = (f_0/m) \exp(\alpha t + i\gamma t)$$

and then take the real part. The result is a forced oscillation of the form

$$x = b \exp(\alpha t) \cos(\gamma t + \delta),$$

where

$$b = f_0/m\sqrt{[(\omega_0^2 + \alpha^2 - \gamma^2 + 2\alpha\lambda)^2 + 4\gamma^2(\alpha + \lambda)^2]},$$
$$\tan \delta = -2\gamma(\alpha + \lambda)/(\omega_0^2 - \gamma^2 + \alpha^2 + 2\alpha\lambda).$$

§27. Parametric resonance

There exist oscillatory systems which are not closed, but in which the external action amounts only to a time variation of the parameters.†

The parameters of a one-dimensional system are the coefficients m and k in the Lagrangian (21.3). If these are functions of time, the equation of motion is

$$\frac{d}{dt}(m\dot{x}) + kx = 0. \tag{27.1}$$

We introduce instead of t a new independent variable τ such that $d\tau = dt/m(t)$; this reduces the equation to

$$d^2x/d\tau^2 + mkx = 0.$$

There is therefore no loss of generality in considering an equation of motion of the form

$$d^2x/dt^2 + \omega^2(t)x = 0 \tag{27.2}$$

obtained from (27.1) if $m = $ constant.

The form of the function $\omega(t)$ is given by the conditions of the problem. Let us assume that this function is periodic with some frequency γ and period $T = 2\pi/\gamma$. This means that $\omega(t + T) = \omega(t)$, and so the equation (27.2) is invariant under the transformation $t \to t + T$. Hence, if $x(t)$ is a solution of the equation, so is $x(t + T)$. That is, if $x_1(t)$ and $x_2(t)$ are two independent integrals of equation (27.2), they must be transformed into linear combinations of themselves when t is replaced by $t + T$. It is possible‡ to choose x_1 and x_2 in such a way that, when $t \to t + T$, they are simply multiplied by

---

† A simple example is that of a pendulum whose point of support executes a given periodic motion in a vertical direction (see Problem 3).

‡ This choice is equivalent to reducing to diagonal form the matrix of the linear transformation of $x_1(t)$ and $x_2(t)$, which involves the solution of the corresponding quadratic secular equation. We shall suppose here that the roots of this equation do not coincide.

constants: $x_1(t+T) = \mu_1 x_1(t)$, $x_2(t+T) = \mu_2 x_2(t)$. The most general functions having this property are

$$x_1(t) = \mu_1{}^{t/T}\Pi_1(t), \qquad x_2(t) = \mu_2{}^{t/T}\Pi_2(t), \tag{27.3}$$

where $\Pi_1(t)$, $\Pi_2(t)$ are purely periodic functions of time with period T.

The constants μ_1 and μ_2 in these functions must be related in a certain way. Multiplying the equations $\ddot{x}_1 + \omega^2(t)x_1 = 0$, $\ddot{x}_2 + \omega^2(t)x_2 = 0$ by x_2 and x_1 respectively and subtracting, we have $\ddot{x}_1 x_2 - \ddot{x}_2 x_1 = \mathrm{d}(\dot{x}_1 x_2 - x_1 \dot{x}_2)\mathrm{d}t = 0$, or

$$\dot{x}_1 x_2 - x_1 \dot{x}_2 = \text{constant.} \tag{27.4}$$

For any functions $x_1(t)$, $x_2(t)$ of the form (27.3), the expression on the left-hand side of (27.4) is multiplied by $\mu_1\mu_2$ when t is replaced by $t+T$. Hence it is clear that, if equation (27.4) is to hold, we must have

$$\mu_1\mu_2 = 1. \tag{27.5}$$

Further information about the constants μ_1, μ_2 can be obtained from the fact that the coefficients in equation (27.2) are real. If $x(t)$ is any integral of such an equation, then the complex conjugate function $x^*(t)$ must also be an integral. Hence it follows that μ_1, μ_2 must be the same as $\mu_1{}^*$, $\mu_2{}^*$, i.e. either $\mu_1 = \mu_2{}^*$ or μ_1 and μ_2 are both real. In the former case, (27.5) gives $\mu_1 = 1/\mu_1{}^*$, i.e. $|\mu_1|^2 = |\mu_2|^2 = 1$: the constants μ_1 and μ_2 are of modulus unity.

In the other case, two independent integrals of equation (27.2) are

$$x_1(t) = \mu^{t/T}\Pi_1(t), \qquad x_2(t) = \mu^{-t/T}\Pi_2(t), \tag{27.6}$$

with a positive or negative real value of μ ($|\mu| \neq 1$). One of these functions (x_1 or x_2 according as $|\mu| > 1$ or $|\mu| < 1$) increases exponentially with time. This means that the system at rest in equilibrium ($x = 0$) is unstable: any deviation from this state, however small, is sufficient to lead to a rapidly increasing displacement x. This is called *parametric resonance*.

It should be noticed that, when the initial values of x and $\dot{x}$ are exactly zero, they remain zero, unlike what happens in ordinary resonance (§22), in which the displacement increases with time (proportionally to t) even from initial values of zero.

Let us determine the conditions for parametric resonance to occur in the important case where the function $\omega(t)$ differs only slightly from a constant value ω_0 and is a simple periodic function:

$$\omega^2(t) = \omega_0{}^2(1 + h \cos\gamma t), \tag{27.7}$$

where the constant $h \ll 1$; we shall suppose h positive, as may always be done by suitably choosing the origin of time. As we shall see below, parametric resonance is strongest if the frequency of the function $\omega(t)$ is nearly twice ω_0. Hence we put $\gamma = 2\omega_0 + \epsilon$, where $\epsilon \ll \omega_0$.

The solution of equation of motion†

$$\ddot{x} + \omega_0^2[1 + h \cos(2\omega_0 + \epsilon)t]x = 0 \tag{27.8}$$

may be sought in the form

$$x = a(t) \cos(\omega_0 + \tfrac{1}{2}\epsilon)t + b(t) \sin(\omega_0 + \tfrac{1}{2}\epsilon)t, \tag{27.9}$$

where $a(t)$ and $b(t)$ are functions of time which vary slowly in comparison with the trigonometrical factors. This form of solution is, of course, not exact. In reality, the function $x(t)$ also involves terms with frequencies which differ from $\omega_0 + \tfrac{1}{2}\epsilon$ by integral multiples of $2\omega_0 + \epsilon$; these terms are, however, of a higher order of smallness with respect to h, and may be neglected in a first approximation (see Problem 1).

We substitute (27.9) in (27.8) and retain only terms of the first order in ϵ, assuming that $\dot{a} \sim \epsilon a$, $\dot{b} \sim \epsilon b$; the correctness of this assumption under resonance conditions is confirmed by the result. The products of trigonometrical functions may be replaced by sums:

$$\cos(\omega_0 + \tfrac{1}{2}\epsilon)t \cdot \cos(2\omega_0 + \epsilon)t = \tfrac{1}{2} \cos 3(\omega_0 + \tfrac{1}{2}\epsilon)t + \tfrac{1}{2} \cos(\omega_0 + \tfrac{1}{2}\epsilon)t,$$

etc., and in accordance with what was said above we omit terms with frequency $3(\omega_0 + \tfrac{1}{2}\epsilon)$. The result is

$$-(2\dot{a} + b\epsilon + \tfrac{1}{2}h\omega_0 b)\omega_0 \sin(\omega_0 + \tfrac{1}{2}\epsilon)t + (2\dot{b} - a\epsilon + \tfrac{1}{2}h\omega_0 a)\omega_0 \cos(\omega_0 + \tfrac{1}{2}\epsilon)t = 0.$$

If this equation is to be justified, the coefficients of the sine and cosine must both be zero. This gives two linear differential equations for the functions $a(t)$ and $b(t)$. As usual, we seek solutions proportional to $\exp(st)$. Then $sa + \tfrac{1}{2}(\epsilon + \tfrac{1}{2}h\omega_0)b = 0$, $\tfrac{1}{2}(\epsilon - \tfrac{1}{2}h\omega_0)a - sb = 0$, and the compatibility condition for these two algebraic equations gives

$$s^2 = \tfrac{1}{4}[(\tfrac{1}{2}h\omega_0)^2 - \epsilon^2]. \tag{27.10}$$

The condition for parametric resonance is that s is real, i.e. $s^2 > 0$.‡ Thus parametric resonance occurs in the range

$$-\tfrac{1}{2}h\omega_0 < \epsilon < \tfrac{1}{2}h\omega_0 \tag{27.11}$$

on either side of the frequency $2\omega_0$.‖ The width of this range is proportional to h, and the values of the amplification coefficient s of the oscillations in the range are of the order of h also.

Parametric resonance also occurs when the frequency γ with which the parameter varies is close to any value $2\omega_0/n$ with n integral. The width of the

---

† An equation of this form (with arbitrary γ and h) is called in mathematical physics *Mathieu's equation.*

‡ The constant μ in (27.6) is related to s by $\mu = -\exp(s\pi/\omega_0)$; when t is replaced by $t + 2\pi/2\omega_0$, the sine and cosine in (27.9) change sign.

‖ If we are interested only in the range of resonance, and not in the values of s in that range, the calculations may be simplified by noting that $s = 0$ at the ends of the range, i.e. the coefficients a and b in (27.9) are constants. This gives immediately $\epsilon = \pm \tfrac{1}{2}h\omega_0$ as in (27.11).

resonance range (region of instability) decreases rapidly with increasing n, however, namely as h^n (see Problem 2, footnote). The amplification coefficient of the oscillations also decreases.

The phenomenon of parametric resonance is maintained in the presence of slight friction, but the region of instability becomes somewhat narrower. As we have seen in §25, friction results in a damping of the amplitude of oscillations as $\exp(-\lambda t)$. Hence the amplification of the oscillations in parametric resonance is as $\exp[(s-\lambda)t]$ with the positive s given by the solution for the frictionless case, and the limit of the region of instability is given by the equation $s-\lambda = 0$. Thus, with s given by (27.10), we have for the resonance range, instead of (27.11),

$$-\sqrt{[(\tfrac{1}{2}h\omega_0)^2 - 4\lambda^2]} < \epsilon < \sqrt{[(\tfrac{1}{2}h\omega_0)^2 - 4\lambda^2]}. \tag{27.12}$$

It should be noticed that resonance is now possible not for arbitrarily small amplitudes h, but only when h exceeds a "threshold" value h_k. When (27.12) holds, $h_k = 4\lambda/\omega_0$. It can be shown that, for resonance near the frequency $2\omega_0/n$, the threshold h_k is proportional to $\lambda^{1/n}$, i.e. it increases with n.

PROBLEMS

PROBLEM 1. Obtain an expression correct as far as the term in h^2 for the limits of the region of instability for resonance near $\gamma = 2\omega_0$.

SOLUTION. We seek the solution of equation (27.8) in the form

$$x = a_0 \cos(\omega_0 + \tfrac{1}{2}\epsilon)t + b_0 \sin(\omega_0 + \tfrac{1}{2}\epsilon)t + a_1 \cos 3(\omega_0 + \tfrac{1}{2}\epsilon)t + b_1 \sin 3(\omega_0 + \tfrac{1}{2}\epsilon)t,$$

which includes terms of one higher order in h than (27.9). Since only the limits of the region of instability are required, we treat the coefficients a_0, b_0, a_1, b_1 as constants in accordance with the last footnote. Substituting in (27.8), we convert the products of trigonometrical functions into sums and omit the terms of frequency $5(\omega_0 + \tfrac{1}{2}\epsilon)$ in this approximation. The result is

$$[-a_0(\omega_0\epsilon + \tfrac{1}{4}\epsilon^2) + \tfrac{1}{2}h\omega_0^2 a_0 + \tfrac{1}{2}h\omega_0^2 a_1] \cos(\omega_0 + \tfrac{1}{2}\epsilon)t +$$

$$+ [-b_0(\omega_0\epsilon + \tfrac{1}{4}\epsilon^2) - \tfrac{1}{2}h\omega_0^2 b_0 + \tfrac{1}{2}h\omega_0^2 b_1] \sin(\omega_0 + \tfrac{1}{2}\epsilon)t +$$

$$+ [\tfrac{1}{2}h\omega_0^2 a_0 - 8\omega_0^2 a_1] \cos 3(\omega_0 + \tfrac{1}{2}\epsilon)t +$$

$$+ [\tfrac{1}{2}h\omega_0^2 b_0 - 8\omega_0^2 b_1] \sin 3(\omega_0 + \tfrac{1}{2}\epsilon)t = 0.$$

In the terms of frequency $\omega_0 + \tfrac{1}{2}\epsilon$ we retain terms of the second order of smallness, but in those of frequency $3(\omega_0 + \tfrac{1}{2}\epsilon)$ only the first-order terms. Each of the expressions in brackets must separately vanish. The last two give $a_1 = ha_0/16$, $b_1 = hb_0/16$, and then the first two give $\omega_0\epsilon \pm \tfrac{1}{2}h\omega_0^2 + \tfrac{1}{4}\epsilon^2 - h^2\omega_0^2/32 = 0$.

Solving this as far as terms of order h^2, we obtain the required limits of ϵ:

$$\epsilon = \pm\tfrac{1}{2}h\omega_0 - h^2\omega_0/32.$$

PROBLEM 2. Determine the limits of the region of instability in resonance near $\gamma = \omega_0$.

SOLUTION. Putting $\gamma = \omega_0 + \epsilon$, we obtain the equation of motion

$$\ddot{x} + \omega_0^2[1 + h\cos(\omega_0 + \epsilon)t]x = 0.$$

Since the required limiting values of $\epsilon \sim h^2$, we seek a solution in the form

$$x = a_0 \cos(\omega_0 + \epsilon)t + b_0 \sin(\omega_0 + \epsilon)t + a_1 \cos 2(\omega_0 + \epsilon)t + b_1 \sin 2(\omega_0 + \epsilon)t + c_1,$$

which includes terms of the first two orders. To determine the limits of instability, we again treat the coefficients as constants, obtaining

$$[-2\omega_0\epsilon a_0 + \tfrac{1}{2}h\omega_0^2 a_1 + h\omega_0^2 c_1]\cos(\omega_0+\epsilon)t +$$
$$+[-2\omega_0\epsilon b_0 + \tfrac{1}{2}h\omega_0^2 b_1]\sin(\omega_0+\epsilon)t +$$
$$+[-3\omega_0^2 a_1 + \tfrac{1}{2}h\omega_0^2 a_0]\cos 2(\omega_0+\epsilon)t +$$
$$+[-3\omega_0^2 b_1 + \tfrac{1}{2}h\omega_0^2 b_0]\sin 2(\omega_0+\epsilon)t + [c_1\omega_0^2 + \tfrac{1}{2}h\omega_0^2 a_0] = 0.$$

Hence $a_1 = ha_0/6$, $b_1 = hb_0/6$, $c_1 = -\tfrac{1}{2}ha_0$, and the limits are† $\epsilon = -5h^2\omega_0/24$, $\epsilon = h^2\omega_0/24$.

PROBLEM 3. Find the conditions for parametric resonance in small oscillations of a simple pendulum whose point of support oscillates vertically.

SOLUTION. The Lagrangian derived in §5, Problem 3(c), gives for small oscillations ($\phi \ll 1$) the equation of motion $\ddot{\phi} + \omega_0^2[1+(4a/l)\cos(2\omega_0+\epsilon)t]\phi = 0$, where $\omega_0^2 = g/l$. Hence we see that the parameter h is here represented by $4a/l$. The condition (27.11), for example, becomes $|\epsilon| < 2a\sqrt{(g/l^3)}$.

§28. Anharmonic oscillations

The whole of the theory of small oscillations discussed above is based on the expansion of the potential and kinetic energies of the system in terms of the co-ordinates and velocities, retaining only the second-order terms. The equations of motion are then linear, and in this approximation we speak of *linear oscillations*. Although such an expansion is entirely legitimate when the amplitude of the oscillations is sufficiently small, in higher approximations (called *anharmonic* or *non-linear oscillations*) some minor but qualitatively different properties of the motion appear.

Let us consider the expansion of the Lagrangian as far as the third-order terms. In the potential energy there appear terms of degree three in the co-ordinates x_i, and in the kinetic energy terms containing products of velocities and co-ordinates, of the form $\dot{x}_i \dot{x}_k x_l$. This difference from the previous expression (23.3) is due to the retention of terms linear in x in the expansion of the functions $a_{ik}(q)$. Thus the Lagrangian is of the form

$$L = \tfrac{1}{2}\sum_{i,k}(m_{ik}\dot{x}_i\dot{x}_k - k_{ik}x_ix_k) +$$
$$+\tfrac{1}{2}\sum_{i,k,l}n_{ikl}\dot{x}_i\dot{x}_k x_l - \tfrac{1}{3}\sum_{i,k,l}l_{ikl}x_ix_kx_l, \qquad (28.1)$$

where n_{ikl}, l_{ikl} are further constant coefficients.

If we change from arbitrary co-ordinates x_i to the normal co-ordinates Q_α of the linear approximation, then, because this transformation is linear, the third and fourth sums in (28.1) become similar sums with Q_α and $\dot{Q}_\alpha$ in place

---

† Generally, the width $\Delta\epsilon$ of the region of instability in resonance near the frequency $2\omega_0/n$ is given by

$$\Delta\epsilon = n^{2n-3}h^n\omega_0/2^{3(n-1)}[(n-1)!]^2,$$

a result due to M. BELL (*Proceedings of the Glasgow Mathematical Association* 3, 132, 1957).

of the co-ordinates x_i and the velocities $\dot{x}_i$. Denoting the coefficients in these new sums by $\lambda_{\alpha\beta\gamma}$ and $\mu_{\alpha\beta\gamma}$, we have the Lagrangian in the form

$$L = \tfrac{1}{2}\sum_\alpha(\dot{Q}_\alpha{}^2 - \omega_\alpha{}^2 Q_\alpha{}^2) + \tfrac{1}{2}\sum_{\alpha,\beta,\gamma}\lambda_{\alpha\beta\gamma}\dot{Q}_\alpha\dot{Q}_\beta Q_\gamma - \tfrac{1}{3}\sum_{\alpha,\beta,\gamma}\mu_{\alpha\beta\gamma}Q_\alpha Q_\beta Q_\gamma. \quad (28.2)$$

We shall not pause to write out in their entirety the equations of motion derived from this Lagrangian. The important feature of these equations is that they are of the form

$$\ddot{Q}_\alpha + \omega_\alpha{}^2 Q_\alpha = f_\alpha(Q, \dot{Q}, \ddot{Q}), \quad (28.3)$$

where f_α are homogeneous functions, of degree two, of the co-ordinates Q and their time derivatives.

Using the method of successive approximations, we seek a solution of these equations in the form

$$Q_\alpha = Q_\alpha{}^{(1)} + Q_\alpha{}^{(2)}, \quad (28.4)$$

where $Q_\alpha{}^{(2)} \ll Q_\alpha{}^{(1)}$, and the $Q_\alpha{}^{(1)}$ satisfy the "unperturbed" equations $\ddot{Q}_\alpha{}^{(1)} + \omega_\alpha{}^2 Q_\alpha{}^{(1)} = 0$, i.e. they are ordinary harmonic oscillations:

$$Q_\alpha{}^{(1)} = a_\alpha \cos(\omega_\alpha t + \alpha_\alpha). \quad (28.5)$$

Retaining only the second-order terms on the right-hand side of (28.3) in the next approximation, we have for the $Q_\alpha{}^{(2)}$ the equations

$$\ddot{Q}_\alpha{}^{(2)} + \omega_\alpha{}^2 Q_\alpha{}^{(2)} = f_\alpha(Q^{(1)}, \dot{Q}^{(1)}, \ddot{Q}^{(1)}), \quad (28.6)$$

where (28.5) is to be substituted on the right. This gives a set of inhomogeneous linear differential equations, in which the right-hand sides can be represented as sums of simple periodic functions. For example,

$$Q_\alpha{}^{(1)}Q_\beta{}^{(1)} = a_\alpha a_\beta \cos(\omega_\alpha t + \alpha_\alpha)\cos(\omega_\beta t + \alpha_\beta)$$
$$= \tfrac{1}{2}a_\alpha a_\beta\{\cos[(\omega_\alpha + \omega_\beta)t + \alpha_\alpha + \alpha_\beta] + \cos[(\omega_\alpha - \omega_\beta)t + \alpha_\alpha - \alpha_\beta]\}.$$

Thus the right-hand sides of equations (28.6) contain terms corresponding to oscillations whose frequencies are the sums and differences of the eigenfrequencies of the system. The solution of these equations must be sought in a form involving similar periodic factors, and so we conclude that, in the second approximation, additional oscillations with frequencies

$$\omega_\alpha \pm \omega_\beta, \quad (28.7)$$

including the double frequencies $2\omega_\alpha$ and the frequency zero (corresponding to a constant displacement), are superposed on the normal oscillations of the system. These are called *combination frequencies*. The corresponding amplitudes are proportional to the products $a_\alpha a_\beta$ (or the squares $a_\alpha{}^2$) of the corresponding normal amplitudes.

In higher approximations, when further terms are included in the expansion of the Lagrangian, combination frequencies occur which are the sums and differences of more than two ω_α; and a further phenomenon also appears.

In the third approximation, the combination frequencies include some which coincide with the original frequencies $\omega_\alpha (= \omega_\alpha + \omega_\beta - \omega_\beta)$. When the method described above is used, the right-hand sides of the equations of motion therefore include resonance terms, which lead to terms in the solution whose amplitude increases with time. It is physically evident, however, that the magnitude of the oscillations cannot increase of itself in a closed system with no external source of energy.

In reality, the fundamental frequencies ω_α in higher approximations are not equal to their "unperturbed" values $\omega_\alpha^{(0)}$ which appear in the quadratic expression for the potential energy. The increasing terms in the solution arise from an expansion of the type

$$\cos(\omega_\alpha^{(0)} + \Delta\omega_\alpha)t \approx \cos \omega_\alpha^{(0)}t - t\Delta\omega_\alpha \sin \omega_\alpha^{(0)}t,$$

which is obviously not legitimate when t is sufficiently large.

In going to higher approximations, therefore, the method of successive approximations must be modified so that the periodic factors in the solution shall contain the exact and not approximate values of the frequencies. The necessary changes in the frequencies are found by solving the equations and requiring that resonance terms should not in fact appear.

We may illustrate this method by taking the example of anharmonic oscillations in one dimension, and writing the Lagrangian in the form

$$L = \tfrac{1}{2}m\dot{x}^2 - \tfrac{1}{2}m\omega_0^2 x^2 - \tfrac{1}{3}m\alpha x^3 - \tfrac{1}{4}m\beta x^4. \tag{28.8}$$

The corresponding equation of motion is

$$\ddot{x} + \omega_0^2 x = -\alpha x^2 - \beta x^3. \tag{28.9}$$

We shall seek the solution as a series of successive approximations: $x = x^{(1)} + x^{(2)} + x^{(3)}$, where

$$x^{(1)} = a \cos \omega t, \tag{28.10}$$

with the exact value of ω, which in turn we express as $\omega = \omega_0 + \omega^{(1)} + \omega^{(2)} + \dots$. (The initial phase in $x^{(1)}$ can always be made zero by a suitable choice of the origin of time.) The form (28.9) of the equation of motion is not the most convenient, since, when (28.10) is substituted in (28.9), the left-hand side is not exactly zero. We therefore rewrite it as

$$\frac{\omega_0^2}{\omega^2}\ddot{x} + \omega_0^2 x = -\alpha x^2 - \beta x^3 - \left(1 - \frac{\omega_0^2}{\omega^2}\right)\ddot{x}. \tag{28.11}$$

Putting $x = x^{(1)} + x^{(2)}$, $\omega = \omega_0 + \omega^{(1)}$ and omitting terms of above the second order of smallness, we obtain for $x^{(2)}$ the equation

$$\begin{aligned}\ddot{x}^{(2)} + \omega_0^2 x^{(2)} &= -\alpha a^2 \cos^2\omega t + 2\omega_0\omega^{(1)}a \cos \omega t \\ &= -\tfrac{1}{2}\alpha a^2 - \tfrac{1}{2}\alpha a^2 \cos 2\omega t + 2\omega_0\omega^{(1)}a \cos \omega t.\end{aligned}$$

The condition for the resonance term to be absent from the right-hand side is simply $\omega^{(1)} = 0$, in agreement with the second approximation discussed at the beginning of this section. Solving the inhomogeneous linear equation in the usual way, we have

$$x^{(2)} = -\frac{\alpha a^2}{2\omega_0^2} + \frac{\alpha a^2}{6\omega_0^2} \cos 2\omega t. \tag{28.12}$$

Putting in (28.11) $x = x^{(1)} + x^{(2)} + x^{(3)}$, $\omega = \omega_0 + \omega^{(2)}$, we obtain the equation for $x^{(3)}$

$$\ddot{x}^{(3)} + \omega_0^2 x^{(3)} = -2\alpha x^{(1)} x^{(2)} - \beta x^{(1)3} + 2\omega_0 \omega^{(2)} x^{(1)}$$

or, substituting on the right-hand side (28.10) and (28.12) and effecting a simple transformation,

$$\ddot{x}^{(3)} + \omega_0^2 x^{(3)} = a^3 \left[\tfrac{1}{4}\beta - \frac{\alpha^2}{6\omega_0^2} \right] \cos 3\omega t +$$

$$+ a \left[2\omega_0\omega^{(2)} + \frac{5a^2\alpha^2}{6\omega_0^2} - \tfrac{3}{4}a^2\beta \right] \cos \omega t.$$

Equating to zero the coefficient of the resonance term $\cos \omega t$, we find the correction to the fundamental frequency, which is proportional to the squared amplitude of the oscillations:

$$\omega^{(2)} = \left(\frac{3\beta}{8\omega_0} - \frac{5\alpha^2}{12\omega_0^3} \right) a^2. \tag{28.13}$$

The combination oscillation of the third order is

$$x^{(3)} = \frac{a^3}{16\omega_0^2} \left(\frac{\alpha^2}{3\omega_0^2} - \tfrac{1}{2}\beta \right) \cos 3\omega t. \tag{28.14}$$

§29. Resonance in non-linear oscillations

When the anharmonic terms in forced oscillations of a system are taken into account, the phenomena of resonance acquire new properties.

Adding to the right-hand side of equation (28.9) an external periodic force of frequency γ, we have

$$\ddot{x} + 2\lambda\dot{x} + \omega_0^2 x = (f/m) \cos \gamma t - \alpha x^2 - \beta x^3; \tag{29.1}$$

here the frictional force, with damping coefficient λ (assumed small) has also been included. Strictly speaking, when non-linear terms are included in the equation of free oscillations, the terms of higher order in the amplitude of the external force (such as occur if it depends on the displacement x) should also be included. We shall omit these terms merely to simplify the formulae; they do not affect the qualitative results.

Let $\gamma = \omega_0 + \epsilon$ with ϵ small, i.e. γ be near the resonance value. To ascertain the resulting type of motion, it is not necessary to consider equation (29.1) if we argue as follows. In the linear approximation, the amplitude b is given near resonance, as a function of the amplitude f and frequency γ of the external force, by formula (26.7), which we write as

$$b^2(\epsilon^2 + \lambda^2) = f^2/4m^2\omega_0{}^2. \tag{29.2}$$

The non-linearity of the oscillations results in the appearance of an amplitude dependence of the eigenfrequency, which we write as

$$\omega_0 + \kappa b^2, \tag{29.3}$$

the constant κ being a definite function of the anharmonic coefficients (see (28.13)). Accordingly, we replace ω_0 by $\omega_0 + \kappa b^2$ in formula (29.2) (or, more precisely, in the small difference $\gamma - \omega_0$). With $\gamma - \omega_0 = \epsilon$, the resulting equation is

$$b^2[(\epsilon - \kappa b^2)^2 + \lambda^2] = f^2/4m^2\omega_0{}^2 \tag{29.4}$$

or

$$\epsilon = \kappa b^2 \pm \sqrt{[(f/2m\omega_0 b)^2 - \lambda^2]}.$$

Equation (29.4) is a cubic equation in b^2, and its real roots give the amplitude of the forced oscillations. Let us consider how this amplitude depends on the frequency of the external force for a given amplitude f of that force.

When f is sufficiently small, the amplitude b is also small, so that powers of b above the second may be neglected in (29.4), and we return to the form of $b(\epsilon)$ given by (29.2), represented by a symmetrical curve with a maximum at the point $\epsilon = 0$ (Fig. 32a). As f increases, the curve changes its shape, though at first it retains its single maximum, which moves to positive ϵ if $\kappa > 0$ (Fig. 32b). At this stage only one of the three roots of equation (29.4) is real.

When f reaches a certain value f_k (to be determined below), however, the nature of the curve changes. For all $f > f_k$ there is a range of frequencies in which equation (29.4) has three real roots, corresponding to the portion $BCDE$ in Fig. 32c.

The limits of this range are determined by the condition $db/d\epsilon = \infty$ which holds at the points D and C. Differentiating equation (29.4) with respect to ϵ, we have

$$db/d\epsilon = (-\epsilon b + \kappa b^3)/(\epsilon^2 + \lambda^2 - 4\kappa\epsilon b^2 + 3\kappa^2 b^4).$$

Hence the points D and C are determined by the simultaneous solution of the equations

$$\epsilon^2 - 4\kappa b^2\epsilon + 3\kappa^2 b^4 + \lambda^2 = 0 \tag{29.5}$$

and (29.4). The corresponding values of ϵ are both positive. The greatest amplitude is reached where $db/d\epsilon = 0$. This gives $\epsilon = \kappa b^2$, and from (29.4) we have

$$b_{\max} = f/2m\omega_0\lambda; \tag{29.6}$$

this is the same as the maximum value given by (29.2).

It may be shown (though we shall not pause to do so here†) that, of the three real roots of equation (29.4), the middle one (represented by the dotted part CD of the curve in Fig. 32c) corresponds to unstable oscillations of the system: any action, no matter how slight, on a system in such a state causes it to oscillate in a manner corresponding to the largest or smallest root (BC or DE). Thus only the branches ABC and DEF correspond to actual oscillations of the system. A remarkable feature here is the existence of a range of frequencies in which two different amplitudes of oscillation are possible. For example, as the frequency of the external force gradually increases, the amplitude of the forced oscillations increases along ABC. At C there is a discontinuity of the amplitude, which falls abruptly to the value corresponding to E, afterwards decreasing along the curve EF as the frequency increases further. If the frequency is now diminished, the amplitude of the forced oscillations varies along FD, afterwards increasing discontinuously from D to B and then decreasing along BA.

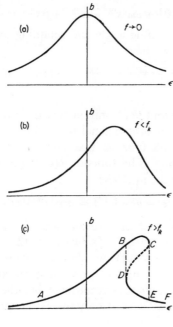

F<small>IG</small>. 32

To calculate the value of f_k, we notice that it is the value of f for which the two roots of the quadratic equation in b^2 (29.5) coincide; for $f = f_k$, the section CD reduces to a point of inflection. Equating to zero the discriminant

---

† The proof is given by, for example, N. N. B<small>OGOLIUBOV</small> and Y. A. M<small>ITROPOLSKY</small>, *Asymptotic Methods in the Theory of Non-Linear Oscillations*, Hindustan Publishing Corporation, Delhi 1961.

of (29.5), we find $\epsilon^2 = 3\lambda^2$, and the corresponding double root is $\kappa b^2 = 2\epsilon/3$. Substitution of these values of b and ϵ in (29.4) gives

$$32m^2\omega_0^2\lambda^3/3\sqrt{3}|\kappa|. \tag{29.7}$$

Besides the change in the nature of the phenomena of resonance at frequencies $\gamma \approx \omega_0$, the non-linearity of the oscillations leads also to new resonances in which oscillations of frequency close to ω_0 are excited by an external force of frequency considerably different from ω_0.

Let the frequency of the external force $\gamma \approx \frac{1}{2}\omega_0$, i.e. $\gamma = \frac{1}{2}\omega_0 + \epsilon$. In the first (linear) approximation, it causes oscillations of the system with the same frequency and with amplitude proportional to that of the force:

$$x^{(1)} = (4f/3m\omega_0^2) \cos(\tfrac{1}{2}\omega_0 + \epsilon)t$$

(see (22.4)). When the non-linear terms are included (second approximation), these oscillations give rise to terms of frequency $2\gamma \approx \omega_0$ on the right-hand side of the equation of motion (29.1). Substituting $x^{(1)}$ in the equation

$$\ddot{x}^{(2)} + 2\lambda\dot{x}^{(2)} + \omega_0^2 x^{(2)} + \alpha x^{(2)2} + \beta x^{(2)3} = -\alpha x^{(1)2},$$

using the cosine of the double angle and retaining only the resonance term on the right-hand side, we have

$$\ddot{x}^{(2)} + 2\lambda\dot{x}^{(2)} + \omega_0^2 x^{(2)} + \alpha x^{(2)2} + \beta x^{(2)3} = -(8\alpha f^2/9m^2\omega_0^4) \cos(\omega_0 + 2\epsilon)t. \tag{29.8}$$

This equation differs from (29.1) only in that the amplitude f of the force is replaced by an expression proportional to f^2. This means that the resulting resonance is of the same type as that considered above for frequencies $\gamma \approx \omega_0$, but is less strong. The function $b(\epsilon)$ is obtained by replacing f by $-8\alpha f^2/9m\omega_0^4$, and ϵ by 2ϵ, in (29.4):

$$b^2[(2\epsilon - \kappa b^2)^2 + \lambda^2] = 16\alpha^2 f^4/81m^4\omega_0^{10}. \tag{29.9}$$

Next, let the frequency of the external force be $\gamma = 2\omega_0 + \epsilon$. In the first approximation, we have $x^{(1)} = -(f/3m\omega_0^2) \cos(2\omega_0 + \epsilon)t$. On substituting $x = x^{(1)} + x^{(2)}$ in equation (29.1), we do not obtain terms representing an external force in resonance such as occurred in the previous case. There is, however, a parametric resonance resulting from the third-order term proportional to the product $x^{(1)}x^{(2)}$. If only this is retained out of the non-linear terms, the equation for $x^{(2)}$ is

$$\ddot{x}^{(2)} + 2\lambda\dot{x}^{(2)} + \omega_0^2 x^{(2)} = -2\alpha x^{(1)}x^{(2)}$$

or

$$\ddot{x}^{(2)} + 2\lambda\dot{x}^{(2)} + \omega_0^2\left[1 - \frac{2\alpha f}{3m\omega_0^4} \cos(2\omega_0 + \epsilon)t\right]x^{(2)} = 0, \tag{29.10}$$

i.e. an equation of the type (27.8) (including friction), which leads, as we have seen, to an instability of the oscillations in a certain range of frequencies.

This equation, however, does not suffice to determine the resulting amplitude of the oscillations. The attainment of a finite amplitude involves non-linear effects, and to include these in the equation of motion we must retain also the terms non-linear in $x^{(2)}$:

$$\ddot{x}^{(2)} + 2\lambda\dot{x}^{(2)} + \omega_0^2 x^{(2)} + \alpha x^{(2)2} + \beta x^{(2)3} = (2\alpha f/3m\omega_0^2)x^{(2)} \cos(2\omega_0 + \epsilon)t. \quad (29.11)$$

The problem can be considerably simplified by virtue of the following fact. Putting on the right-hand side of (29.11) $x^{(2)} = b \cos[(\omega_0 + \frac{1}{2}\epsilon)t + \delta]$, where b is the required amplitude of the resonance oscillations and δ a constant phase difference which is of no importance in what follows, and writing the product of cosines as a sum, we obtain a term $(\alpha f b/3m\omega_0^2) \cos[(\omega_0 + \frac{1}{2}\epsilon)t - \delta]$ of the ordinary resonance type (with respect to the eigenfrequency ω_0 of the system). The problem thus reduces to that considered at the beginning of this section, namely ordinary resonance in a non-linear system, the only differences being that the amplitude of the external force is here represented by $\alpha f b/3\omega_0^2$, and ϵ is replaced by $\frac{1}{2}\epsilon$. Making this change in equation (29.4), we have

$$b^2[(\tfrac{1}{2}\epsilon - \kappa b^2)^2 + \lambda^2] = \alpha^2 f^2 b^2/36m^2\omega_0^6.$$

Solving for b, we find the possible values of the amplitude:

$$b = 0, \quad (29.12)$$

$$b^2 = \frac{1}{\kappa}[\tfrac{1}{2}\epsilon + \sqrt{\{(\alpha f/6m\omega_0^3)^2 - \lambda^2\}}], \quad (29.13)$$

$$b^2 = \frac{1}{\kappa}[\tfrac{1}{2}\epsilon - \sqrt{\{(\alpha f/6m\omega_0^3)^2 - \lambda^2\}}]. \quad (29.14)$$

Figure 33 shows the resulting dependence of b on ϵ for $\kappa > 0$; for $\kappa < 0$ the curves are the reflections (in the b-axis) of those shown. The points B and C correspond to the values $\epsilon = \pm\sqrt{\{(\alpha f/3m\omega_0^3)^2 - 4\lambda^2\}}$. To the left of B, only the value $b = 0$ is possible, i.e. there is no resonance, and oscillations of frequency near ω_0 are not excited. Between B and C there are two roots, $b = 0 \, (BC)$ and (29.13) (BE). Finally, to the right of C there are three roots (29.12)–(29.14). Not all these, however, correspond to stable oscillations. The value $b = 0$ is unstable on BC,† and it can also be shown that the middle root (29.14) always gives instability. The unstable values of b are shown in Fig. 33 by dashed lines.

Let us examine, for example, the behaviour of a system initially "at rest"‡ as the frequency of the external force is gradually diminished. Until the point

---

† This segment corresponds to the region of parametric resonance (27.12), and a comparison of (29.10) and (27.8) gives $|h| = 2\alpha f/3m\omega_0^4$. The condition $|2\alpha f/3m\omega_0^3| > 4\lambda$ for which the phenomenon can exist corresponds to $h > h_k$.

‡ It should be recalled that only resonance phenomena are under consideration. If these phenomena are absent, the system is not literally at rest, but executes small forced oscillations of frequency γ.

C is reached, $b = 0$, but at C the state of the system passes discontinuously to the branch EB. As ϵ decreases further, the amplitude of the oscillations decreases to zero at B. When the frequency increases again, the amplitude increases along BE.†

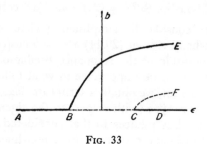

FIG. 33

The cases of resonance discussed above are the principal ones which may occur in a non-linear oscillating system. In higher approximations, resonances appear at other frequencies also. Strictly speaking, a resonance must occur at every frequency γ for which $n\gamma + m\omega_0 = \omega_0$ with n and m integers, i.e. for every $\gamma = p\omega_0/q$ with p and q integers. As the degree of approximation increases, however, the strength of the resonances, and the widths of the frequency ranges in which they occur, decrease so rapidly that in practice only the resonances at frequencies $\gamma \approx p\omega_0/q$ with small p and q can be observed.

PROBLEM

Determine the function $b(\epsilon)$ for resonance at frequencies $\gamma \approx 3\omega_0$.

SOLUTION. In the first approximation, $x^{(1)} = -(f/8m\omega_0^2) \cos(3\omega_0+\epsilon)t$. For the second approximation $x^{(2)}$ we have from (29.1) the equation

$$\ddot{x}^{(2)} + 2\lambda\dot{x}^{(2)} + \omega_0^2 x^{(2)} + \alpha x^{(2)2} + \beta x^{(2)3} = -3\beta x^{(1)}x^{(2)2},$$

where only the term which gives the required resonance has been retained on the right-hand side. Putting $x^{(2)} = b \cos[(\omega_0+\tfrac{1}{3}\epsilon)t+\delta]$ and taking the resonance term out of the product of three cosines, we obtain on the right-hand side the expression

$$(3\beta b^2 f/32m\omega_0^2) \cos[(\omega_0+\tfrac{1}{3}\epsilon)t-2\delta].$$

Hence it is evident that $b(\epsilon)$ is obtained by replacing f by $3\beta b^2 f/32\omega_0^2$, and ϵ by $\tfrac{1}{3}\epsilon$, in (29.4):

$$b^2[(\tfrac{1}{3}\epsilon - \kappa b^2)^2 + \lambda^2] = (9\beta^2 f^2/2^{12}m^2\omega_0^6)b^4 \equiv Ab^4.$$

The roots of this equation are

$$b = 0, \qquad b^2 = \frac{\epsilon}{3\kappa} + \frac{A}{2\kappa^2} \pm \frac{1}{\kappa}\sqrt{\left(\frac{\epsilon A}{3\kappa} + \frac{A^2}{4\kappa^2} - \lambda^2\right)}.$$

Fig. 34 shows a graph of the function $b(\epsilon)$ for $\kappa > 0$. Only the value $b = 0$ (the ϵ-axis) and the branch AB corresponds to stability. The point A corresponds to $\epsilon_k = 3(4\kappa^2\lambda^2 - A^2)/4\kappa A$,

---

† It must be noticed, however, that all the formulae derived here are valid only when the amplitude b (and also ϵ) is sufficiently small. In reality, the curves BE and CF meet, and at their point of intersection the oscillation ceases; thereafter, $b = 0$.

$b_k{}^2 = (4\kappa^2\lambda^2 + A^2)/4\kappa^2 A$. Oscillations exist only for $\epsilon > \epsilon_k$, and then $b > b_k$. Since the state $b = 0$ is always stable, an initial "push" is necessary in order to excite oscillations.

The formulae given above are valid only for small ϵ. This condition is satisfied if λ is small and also the amplitude of the force is such that $\kappa\lambda^2/\omega_0 \ll A \ll \kappa\omega_0$.

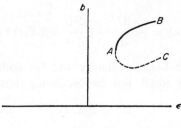

FIG. 34

§30. Motion in a rapidly oscillating field

Let us consider the motion of a particle subject both to a time-independent field of potential U and to a force

$$f = f_1 \cos \omega t + f_2 \sin \omega t \qquad (30.1)$$

which varies in time with a high frequency ω (f_1, f_2 being functions of the co-ordinates only). By a "high" frequency we mean one such that $\omega \gg 1/T$, where T is the order of magnitude of the period of the motion which the particle would execute in the field U alone. The magnitude of f is not assumed small in comparison with the forces due to the field U, but we shall assume that the oscillation (denoted below by ξ) of the particle as a result of this force is small.

To simplify the calculations, let us first consider motion in one dimension in a field depending only on the space co-ordinate x. Then the equation of motion of the particle is†

$$m\ddot{x} = -dU/dx + f. \qquad (30.2)$$

It is evident, from the nature of the field in which the particle moves, that it will traverse a smooth path and at the same time execute small oscillations of frequency ω about that path. Accordingly, we represent the function $x(t)$ as a sum:

$$x(t) = X(t) + \xi(t), \qquad (30.3)$$

where $\xi(t)$ corresponds to these small oscillations.

The mean value of the function $\xi(t)$ over its period $2\pi/\omega$ is zero, and the function $X(t)$ changes only slightly in that time. Denoting this average by a bar, we therefore have $\bar{x} = X(t)$, i.e. $X(t)$ describes the "smooth" motion of

---

† The co-ordinate x need not be Cartesian, and the coefficient m is therefore not necessarily the mass of the particle, nor need it be constant as has been assumed in (30.2). This assumption, however, does not affect the final result (see the last footnote to this section).

the particle averaged over the rapid oscillations. We shall derive an equation which determines the function $X(t)$.†

Substituting (30.3) in (30.2) and expanding in powers of ξ as far as the first-order terms, we obtain

$$m\ddot{X} + m\ddot{\xi} = -\frac{\mathrm{d}U}{\mathrm{d}x} - \xi\frac{\mathrm{d}^2U}{\mathrm{d}x^2} + f(X, t) + \xi\frac{\partial f}{\partial X}. \tag{30.4}$$

This equation involves both oscillatory and "smooth" terms, which must evidently be separately equal. For the oscillating terms we can put simply

$$m\ddot{\xi} = f(X, t); \tag{30.5}$$

the other terms contain the small factor ξ and are therefore of a higher order of smallness (but the derivative $\ddot{\xi}$ is proportional to the large quantity ω^2 and so is not small). Integrating equation (30.5) with the function f given by (30.1) (regarding X as a constant), we have

$$\xi = -f/m\omega^2. \tag{30.6}$$

Next, we average equation (30.4) with respect to time (in the sense discussed above). Since the mean values of the first powers of f and ξ are zero, the result is

$$m\ddot{X} = -\frac{\mathrm{d}U}{\mathrm{d}X} + \overline{\xi\frac{\partial f}{\partial X}} = -\frac{\mathrm{d}U}{\mathrm{d}X} - \frac{1}{m\omega^2}\overline{f\frac{\partial f}{\partial X}},$$

which involves only the function $X(t)$. This equation can be written

$$m\ddot{X} = -\mathrm{d}U_{\text{eff}}/\mathrm{d}X, \tag{30.7}$$

where the "effective potential energy" is defined as‡

$$U_{\text{eff}} = U + \overline{f^2}/2m\omega^2$$
$$= U + (f_1^2 + f_2^2)/4m\omega^2. \tag{30.8}$$

Comparing this expression with (30.6), we easily see that the term added to U is just the mean kinetic energy of the oscillatory motion:

$$U_{\text{eff}} = U + \tfrac{1}{2}m\overline{\dot{\xi}^2}. \tag{30.9}$$

Thus the motion of the particle averaged over the oscillations is the same as if the constant potential U were augmented by a constant quantity proportional to the squared amplitude of the variable field.

_____

† The principle of this derivation is due to P. L. KAPITZA (1951).
‡ By means of somewhat more lengthy calculations it is easy to show that formulae (30.7) and (30.8) remain valid even if m is a function of x.

The result can easily be generalised to the case of a system with any number of degrees of freedom, described by generalised co-ordinates q_i. The effective potential energy is then given not by (30.8), but by

$$U_{\text{eff}} = U + \frac{1}{2\omega^2} \sum_{i,k} a^{-1}{}_{ik} \overline{f_i f_k}$$

$$= U + \sum_{i,k} \tfrac{1}{2} a_{ik} \overline{\dot{\xi}_i \dot{\xi}_k}, \tag{30.10}$$

where the quantities $a^{-1}{}_{ik}$, which are in general functions of the co-ordinates, are the elements of the matrix inverse to the matrix of the coefficients a_{ik} in the kinetic energy (5.5) of the system.

PROBLEMS

PROBLEM 1. Determine the positions of stable equilibrium of a pendulum whose point of support oscillates vertically with a high frequency γ ($\gg \sqrt{(g/l)}$).

SOLUTION. From the Lagrangian derived in §5, Problem 3(c), we see that in this case the variable force is $f = -mla\gamma^2 \cos \gamma t \sin \phi$ (the quantity x being here represented by the angle ϕ). The "effective potential energy" is therefore $U_{\text{eff}} = mgl[-\cos \phi + (a^2\gamma^2/4gl) \sin^2\phi]$. The positions of stable equilibrium correspond to the minima of this function. The vertically downward position ($\phi = 0$) is always stable. If the condition $a^2\gamma^2 > 2gl$ holds, the vertically upward position ($\phi = \pi$) is also stable.

PROBLEM 2. The same as Problem 1, but for a pendulum whose point of support oscillates horizontally.

SOLUTION. From the Lagrangian derived in §5, Problem 3(b), we find $f = mla\gamma^2 \cos \gamma t \cos \phi$ and $U_{\text{eff}} = mgl[-\cos \phi + (a^2\gamma^2/4gl) \cos^2\phi]$. If $a^2\gamma^2 < 2gl$, the position $\phi = 0$ is stable. If $a^2\gamma^2 > 2gl$, on the other hand, the stable equilibrium position is given by $\cos \phi = 2gl/a^2\gamma^2$.

CHAPTER VI

MOTION OF A RIGID BODY

§31. Angular velocity

A *rigid body* may be defined in mechanics as a system of particles such that the distances between the particles do not vary. This condition can, of course, be satisfied only approximately by systems which actually exist in nature. The majority of solid bodies, however, change so little in shape and size under ordinary conditions that these changes may be entirely neglected in considering the laws of motion of the body as a whole.

In what follows, we shall often simplify the derivations by regarding a rigid body as a discrete set of particles, but this in no way invalidates the assertion that solid bodies may usually be regarded in mechanics as continuous, and their internal structure disregarded. The passage from the formulae which involve a summation over discrete particles to those for a continuous body is effected by simply replacing the mass of each particle by the mass $\rho \, dV$ contained in a volume element dV (ρ being the density) and the summation by an integration over the volume of the body.

To describe the motion of a rigid body, we use two systems of co-ordinates: a "fixed" (i.e. inertial) system XYZ, and a moving system $x_1 = x$, $x_2 = y$, $x_3 = z$ which is supposed to be rigidly fixed in the body and to participate in its motion. The origin of the moving system may conveniently be taken to coincide with the centre of mass of the body.

The position of the body with respect to the fixed system of co-ordinates is completely determined if the position of the moving system is specified. Let the origin O of the moving system have the radius vector $\mathbf{R}$ (Fig. 35). The orientation of the axes of that system relative to the fixed system is given by three independent angles, which together with the three components of the vector $\mathbf{R}$ make six co-ordinates. Thus a rigid body is a mechanical system with six degrees of freedom.

Let us consider an arbitrary infinitesimal displacement of a rigid body. It can be represented as the sum of two parts. One of these is an infinitesimal translation of the body, whereby the centre of mass moves to its final position, but the orientation of the axes of the moving system of co-ordinates is unchanged. The other is an infinitesimal rotation about the centre of mass, whereby the remainder of the body moves to its final position.

Let $\mathbf{r}$ be the radius vector of an arbitrary point P in a rigid body in the moving system, and $\mathbf{t}$ the radius vector of the same point in the fixed system (Fig. 35). Then the infinitesimal displacement $d\mathbf{t}$ of P consists of a displacement $d\mathbf{R}$, equal to that of the centre of mass, and a displacement $d\boldsymbol{\phi} \times \mathbf{r}$

relative to the centre of mass resulting from a rotation through an infinitesimal angle $d\phi$ (see (9.1)): $d\mathbf{r} = d\mathbf{R} + d\boldsymbol{\phi} \times \mathbf{r}$. Dividing this equation by the time dt during which the displacement occurs, and putting†

$$d\mathbf{r}/dt = \mathbf{v}, \qquad d\mathbf{R}/dt = \mathbf{V}, \qquad d\boldsymbol{\phi}/dt = \boldsymbol{\Omega}, \tag{31.1}$$

we obtain the relation

$$\mathbf{v} = \mathbf{V} + \boldsymbol{\Omega} \times \mathbf{r}. \tag{31.2}$$

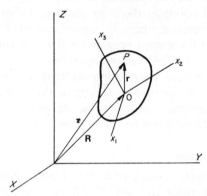

FIG. 35

The vector $\mathbf{V}$ is the velocity of the centre of mass of the body, and is also the *translational velocity* of the body. The vector $\boldsymbol{\Omega}$ is called the *angular velocity* of the rotation of the body; its direction, like that of $d\boldsymbol{\phi}$, is along the axis of rotation. Thus the velocity $\mathbf{v}$ of any point in the body relative to the fixed system of co-ordinates can be expressed in terms of the translational velocity of the body and its angular velocity of rotation.

It should be emphasised that, in deriving formula (31.2), no use has been made of the fact that the origin is located at the centre of mass. The advantages of this choice of origin will become evident when we come to calculate the energy of the moving body.

Let us now assume that the system of co-ordinates fixed in the body is such that its origin is not at the centre of mass O, but at some point O' at a distance $\mathbf{a}$ from O. Let the velocity of O' be $\mathbf{V}'$, and the angular velocity of the new system of co-ordinates be $\boldsymbol{\Omega}'$. We again consider some point P in the body, and denote by $\mathbf{r}'$ its radius vector with respect to O'. Then $\mathbf{r} = \mathbf{r}' + \mathbf{a}$, and substitution in (31.2) gives $\mathbf{v} = \mathbf{V} + \boldsymbol{\Omega} \times \mathbf{a} + \boldsymbol{\Omega} \times \mathbf{r}'$. The definition of $\mathbf{V}'$ and $\boldsymbol{\Omega}'$ shows that $\mathbf{v} = \mathbf{V}' + \boldsymbol{\Omega}' \times \mathbf{r}'$. Hence it follows that

$$\mathbf{V}' = \mathbf{V} + \boldsymbol{\Omega} \times \mathbf{a}, \qquad \boldsymbol{\Omega}' = \boldsymbol{\Omega}. \tag{31.3}$$

The second of these equations is very important. We see that the angular velocity of rotation, at any instant, of a system of co-ordinates fixed in the body is independent of the particular system chosen. All such systems

---

† To avoid any misunderstanding, it should be noted that this way of expressing the angular velocity is somewhat arbitrary: the vector $\delta\boldsymbol{\phi}$ exists only for an infinitesimal rotation, and not for all finite rotations.

rotate with angular velocities $\boldsymbol{\Omega}$ which are equal in magnitude and parallel in direction. This enables us to call $\boldsymbol{\Omega}$ the *angular velocity of the body*. The velocity of the translational motion, however, does not have this "absolute" property.

It is seen from the first formula (31.3) that, if $\mathbf{V}$ and $\boldsymbol{\Omega}$ are, at any given instant, perpendicular for some choice of the origin O, then $\mathbf{V}'$ and $\boldsymbol{\Omega}'$ are perpendicular for any other origin O'. Formula (31.2) shows that in this case the velocities $\mathbf{v}$ of all points in the body are perpendicular to $\boldsymbol{\Omega}$. It is then always possible† to choose an origin O' whose velocity $\mathbf{V}'$ is zero, so that the motion of the body at the instant considered is a pure rotation about an axis through O'. This axis is called the *instantaneous axis of rotation*.‡

In what follows we shall always suppose that the origin of the moving system is taken to be at the centre of mass of the body, and so the axis of rotation passes through the centre of mass. In general both the magnitude and the direction of $\boldsymbol{\Omega}$ vary during the motion.

§32. The inertia tensor

To calculate the kinetic energy of a rigid body, we may consider it as a discrete system of particles and put $T = \sum \frac{1}{2}mv^2$, where the summation is taken over all the particles in the body. Here, and in what follows, we simplify the notation by omitting the suffix which denumerates the particles.

Substitution of (31.2) gives

$$T = \sum \tfrac{1}{2}m(\mathbf{V} + \boldsymbol{\Omega} \times \mathbf{r})^2 = \sum \tfrac{1}{2}mV^2 + \sum m\mathbf{V} \cdot \boldsymbol{\Omega} \times \mathbf{r} + \sum \tfrac{1}{2}m(\boldsymbol{\Omega} \times \mathbf{r})^2.$$

The velocities $\mathbf{V}$ and $\boldsymbol{\Omega}$ are the same for every point in the body. In the first term, therefore, $\frac{1}{2}V^2$ can be taken outside the summation sign, and $\sum m$ is just the mass of the body, which we denote by μ. In the second term we put $\sum m\mathbf{V} \cdot \boldsymbol{\Omega} \times \mathbf{r} = \sum m\mathbf{r} \cdot \mathbf{V} \times \boldsymbol{\Omega} = \mathbf{V} \times \boldsymbol{\Omega} \cdot \sum m\mathbf{r}$. Since we take the origin of the moving system to be at the centre of mass, this term is zero, because $\sum m\mathbf{r} = 0$. Finally, in the third term we expand the squared vector product. The result is

$$T = \tfrac{1}{2}\mu V^2 + \tfrac{1}{2} \sum m[\Omega^2 r^2 - (\boldsymbol{\Omega} \cdot \mathbf{r})^2]. \tag{32.1}$$

Thus the kinetic energy of a rigid body can be written as the sum of two parts. The first term in (32.1) is the kinetic energy of the translational motion, and is of the same form as if the whole mass of the body were concentrated at the centre of mass. The second term is the kinetic energy of the rotation with angular velocity $\boldsymbol{\Omega}$ about an axis passing through the centre of mass. It should be emphasised that this division of the kinetic energy into two parts is possible only because the origin of the co-ordinate system fixed in the body has been taken to be at its centre of mass.

---

† O' may, of course, lie outside the body.

‡ In the general case where $\mathbf{V}$ and $\boldsymbol{\Omega}$ are not perpendicular, the origin may be chosen so as to make $\mathbf{V}$ and $\boldsymbol{\Omega}$ parallel, i.e. so that the motion consists (at the instant in question) of a rotation about some axis together with a translation along that axis.

We may rewrite the kinetic energy of rotation in tensor form, i.e. in terms of the components† x_i and Ω_i of the vectors $\mathbf{r}$ and $\boldsymbol{\Omega}$. We have

$$
\begin{aligned}
T_{\text{rot}} &= \tfrac{1}{2} \sum m(\Omega_i^2 x_i^2 - \Omega_i x_i \Omega_k x_k) \\
&= \tfrac{1}{2} \sum m(\Omega_i \Omega_k \delta_{ik} x_l^2 - \Omega_i \Omega_k x_i x_k) \\
&= \tfrac{1}{2} \Omega_i \Omega_k \sum m(x_l^2 \delta_{ik} - x_i x_k).
\end{aligned}
$$

Here we have used the identity $\Omega_i = \delta_{ik}\Omega_k$, where δ_{ik} is the unit tensor, whose components are unity for $i = k$ and zero for $i \neq k$. In terms of the tensor

$$
I_{ik} = \sum m(x_l^2 \delta_{ik} - x_i x_k) \tag{32.2}
$$

we have finally the following expression for the kinetic energy of a rigid body:

$$
T = \tfrac{1}{2}\mu V^2 + \tfrac{1}{2} I_{ik}\Omega_i \Omega_k. \tag{32.3}
$$

The Lagrangian for a rigid body is obtained from (32.3) by subtracting the potential energy:

$$
L = \tfrac{1}{2}\mu V^2 + \tfrac{1}{2} I_{ik}\Omega_i \Omega_k - U. \tag{32.4}
$$

The potential energy is in general a function of the six variables which define the position of the rigid body, e.g. the three co-ordinates X, Y, Z of the centre of mass and the three angles which specify the relative orientation of the moving and fixed co-ordinate axes.

The tensor I_{ik} is called the *inertia tensor* of the body. It is symmetrical, i.e.

$$
I_{ik} = I_{ki}, \tag{32.5}
$$

as is evident from the definition (32.2). For clarity, we may give its components explicitly:

$$
I_{ik} = \begin{bmatrix}
\sum m(y^2 + z^2) & -\sum mxy & -\sum mxz \\
-\sum myx & \sum m(x^2 + z^2) & -\sum myz \\
-\sum mzx & -\sum mzy & \sum m(x^2 + y^2)
\end{bmatrix}. \tag{32.6}
$$

The components I_{xx}, I_{yy}, I_{zz} are called the *moments of inertia* about the corresponding axes.

The inertia tensor is evidently additive: the moments of inertia of a body are the sums of those of its parts.

---

† In this chapter, the letters i, k, l are tensor suffixes and take the values 1, 2, 3. The summation rule will always be used, i.e. summation signs are omitted, but summation over the values 1, 2, 3 is implied whenever a suffix occurs twice in any expression. Such a suffix is called a *dummy suffix*. For example, $A_i B_i = \mathbf{A} \cdot \mathbf{B}$, $A_i^2 = A_i A_i = \mathbf{A}^2$, etc. It is obvious that dummy suffixes can be replaced by any other like suffixes, except ones which already appear elsewhere in the expression concerned.

If the body is regarded as continuous, the sum in the definition (32.2) becomes an integral over the volume of the body:

$$I_{ik} = \int \rho(x_l^2 \delta_{ik} - x_i x_k) \, dV. \tag{32.7}$$

Like any symmetrical tensor of rank two, the inertia tensor can be reduced to diagonal form by an appropriate choice of the directions of the axes x_1, x_2, x_3. These directions are called the *principal axes of inertia*, and the corresponding values of the diagonal components of the tensor are called the *principal moments of inertia*; we shall denote them by I_1, I_2, I_3. When the axes x_1, x_2, x_3 are so chosen, the kinetic energy of rotation takes the very simple form

$$T_{\text{rot}} = \tfrac{1}{2}(I_1\Omega_1^2 + I_2\Omega_2^2 + I_3\Omega_3^2). \tag{32.8}$$

None of the three principal moments of inertia can exceed the sum of the other two. For instance,

$$I_1 + I_2 = \sum m(x_1^2 + x_2^2 + 2x_3^2) \geqslant \sum m(x_1^2 + x_2^2) = I_3. \tag{32.9}$$

A body whose three principal moments of inertia are all different is called an *asymmetrical top*. If two are equal ($I_1 = I_2 \neq I_3$), we have a *symmetrical top*. In this case the direction of one of the principal axes in the x_1x_2-plane may be chosen arbitrarily. If all three principal moments of inertia are equal, the body is called a *spherical top*, and the three axes of inertia may be chosen arbitrarily as any three mutually perpendicular axes.

The determination of the principal axes of inertia is much simplified if the body is symmetrical, for it is clear that the position of the centre of mass and the directions of the principal axes must have the same symmetry as the body. For example, if the body has a plane of symmetry, the centre of mass must lie in that plane, which also contains two of the principal axes of inertia, while the third is perpendicular to the plane. An obvious case of this kind is a coplanar system of particles. Here there is a simple relation between the three principal moments of inertia. If the plane of the system is taken as the x_1x_2-plane, then $x_3 = 0$ for every particle, and so $I_1 = \Sigma mx_2^2$, $I_2 = \Sigma mx_1^2$, $I_3 = \Sigma m(x_1^2 + x_2^2)$, whence

$$I_3 = I_1 + I_2. \tag{32.10}$$

If a body has an axis of symmetry of any order, the centre of mass must lie on that axis, which is also one of the principal axes of inertia, while the other two are perpendicular to it. If the axis is of order higher than the second, the body is a symmetrical top. For any principal axis perpendicular to the axis of symmetry can be turned through an angle different from 180° about the latter, i.e. the choice of the perpendicular axes is not unique, and this can happen only if the body is a symmetrical top.

A particular case here is a collinear system of particles. If the line of the system is taken as the x_3-axis, then $x_1 = x_2 = 0$ for every particle, and so two of the principal moments of inertia are equal and the third is zero:

$$I_1 = I_2 = \sum mx_3^2, \qquad I_3 = 0. \qquad (32.11)$$

Such a system is called a *rotator*. The characteristic property which distinguishes a rotator from other bodies is that it has only two, not three, rotational degrees of freedom, corresponding to rotations about the x_1 and x_2 axes: it is clearly meaningless to speak of the rotation of a straight line about itself.

Finally, we may note one further result concerning the calculation of the inertia tensor. Although this tensor has been defined with respect to a system of co-ordinates whose origin is at the centre of mass (as is necessary if the fundamental formula (32.3) is to be valid), it may sometimes be more conveniently found by first calculating a similar tensor $I'_{ik} = \sum m(x'^2_i \delta_{ik} - x'_i x'_k)$, defined with respect to some other origin O'. If the distance OO' is represented by a vector $\mathbf{a}$, then $\mathbf{r} = \mathbf{r}' + \mathbf{a}$, $x_i = x'_i + a_i$; since, by the definition of O, $\sum m\mathbf{r} = 0$, we have

$$I'_{ik} = I_{ik} + \mu(a^2\delta_{ik} - a_i a_k). \qquad (32.12)$$

Using this formula, we can easily calculate I_{ik} if I'_{ik} is known.

PROBLEMS

PROBLEM 1. Determine the principal moments of inertia for the following types of molecule, regarded as systems of particles at fixed distances apart: (a) a molecule of collinear atoms, (b) a triatomic molecule which is an isosceles triangle (Fig. 36), (c) a tetratomic molecule which is an equilateral-based tetrahedron (Fig. 37).

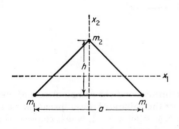

FIG. 36

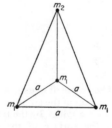

FIG. 37

SOLUTION. (a)

$$I_1 = I_2 = \frac{1}{\mu} \sum_{a \neq b} m_a m_b l_{ab}^2, \qquad I_3 = 0,$$

where m_a is the mass of the ath atom, l_{ab} the distance between the ath and bth atoms, and the summation includes one term for every pair of atoms in the molecule.

For a diatomic molecule there is only one term in the sum, and the result is obvious: it is the product of the reduced mass of the two atoms and the square of the distance between them: $I_1 = I_2 = m_1 m_2 l^2/(m_1 + m_2)$.

(b) The centre of mass is on the axis of symmetry of the triangle, at a distance $X_2 = m_2 h/\mu$ from its base (h being the height of the triangle). The moments of inertia are $I_1 = 2m_1 m_2 h^2/\mu$, $I_2 = \frac{1}{2}m_1 a^2$, $I_3 = I_1 + I_2$.

(c) The centre of mass is on the axis of symmetry of the tetrahedron, at a distance $X_3 = m_2 h/\mu$ from its base (h being the height of the tetrahedron). The moments of inertia are $I_1 = I_2 = 3m_1m_2h^2/\mu + \frac{1}{2}m_1a^2$, $I_3 = m_1a^2$. If $m_1 = m_2$, $h = \surd(2/3)a$, the molecule is a regular tetrahedron and $I_1 = I_2 = I_3 = m_1a^2$.

PROBLEM 2. Determine the principal moments of inertia for the following homogeneous bodies: (a) a thin rod of length l, (b) a sphere of radius R, (c) a circular cylinder of radius R and height h, (d) a rectangular parallelepiped of sides a, b, and c, (e) a circular cone of height h and base radius R, (f) an ellipsoid of semiaxes a, b, c.

SOLUTION. (a) $I_1 = I_2 = \frac{1}{12}\mu l^2$, $I_3 = 0$ (we neglect the thickness of the rod).
(b) $I_1 = I_2 = I_3 = \frac{2}{5}\mu R^2$ (found by calculating the sum $I_1+I_2+I_3 = 2\rho \int r^2 \, dV$).
(c) $I_1 = I_2 = \frac{1}{4}\mu(R^2+\frac{1}{3}h^2)$, $I_3 = \frac{1}{2}\mu R^2$ (where the x_3-axis is along the axis of the cylinder).
(d) $I_1 = \frac{1}{12}\mu(b^2+c^2)$, $I_2 = \frac{1}{12}\mu(a^2+c^2)$, $I_3 = \frac{1}{12}\mu(a^2+b^2)$ (where the axes x_1, x_2, x_3 are along the sides a, b, c respectively).
(e) We first calculate the tensor I'_{ik} with respect to axes whose origin is at the vertex of the cone (Fig. 38). The calculation is simple if cylindrical co-ordinates are used, and the result is $I'_1 = I'_2 = \frac{3}{5}\mu(\frac{1}{4}R^2+h^2)$, $I'_3 = \frac{3}{10}\mu R^2$. The centre of mass is easily shown to be on the axis of the cone and at a distance $a = \frac{3}{4}h$ from the vertex. Formula (32.12) therefore gives $I_1 = I_2 = I'_1 - \mu a^2 = \frac{3}{20}\mu(R^2+\frac{1}{4}h^2)$, $I_3 = I'_3 = \frac{3}{10}\mu R^2$.

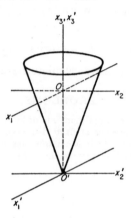

FIG. 38

(f) The centre of mass is at the centre of the ellipsoid, and the principal axes of inertia are along the axes of the ellipsoid. The integration over the volume of the ellipsoid can be reduced to one over a sphere by the transformation $x = a\xi$, $y = b\eta$, $z = c\zeta$, which converts the equation of the surface of the ellipsoid $x^2/a^2+y^2/b^2+z^2/c^2 = 1$ into that of the unit sphere $\xi^2+\eta^2+\zeta^2 = 1$.

For example, the moment of inertia about the x-axis is

$$I_1 = \rho \iiint (y^2+z^2) \, dx \, dy \, dz$$
$$= \rho \, abc \iiint (b^2\eta^2+c^2\zeta^2) \, d\xi \, d\eta \, d\zeta$$
$$= \frac{1}{5}abcI'(b^2+c^2),$$

where I' is the moment of inertia of a sphere of unit radius. Since the volume of the ellipsoid is $4\pi abc/3$, we find the moments of inertia $I_1 = \frac{1}{5}\mu(b^2+c^2)$, $I_2 = \frac{1}{5}\mu(a^2+c^2)$, $I_3 = \frac{1}{5}\mu(a^2+b^2)$.

PROBLEM 3. Determine the frequency of small oscillations of a compound pendulum (a rigid body swinging about a fixed horizontal axis in a gravitational field).

SOLUTION. Let l be the distance between the centre of mass of the pendulum and the axis about which it rotates, and α, β, γ the angles between the principal axes of inertia and the axis of rotation. We take as the variable co-ordinate the angle ϕ between the vertical and a line through the centre of mass perpendicular to the axis of rotation. The velocity of the centre of mass is $V = l\dot\phi$, and the components of the angular velocity along the principal axes of inertia are $\phi \cos \alpha$, $\phi \cos \beta$, $\phi \cos \gamma$. Assuming the angle ϕ to be small, we find the potential energy $U = \mu g l(1 - \cos \phi) \approx \frac{1}{2}\mu g l \phi^2$. The Lagrangian is therefore

$$L = \tfrac{1}{2}\mu l^2 \dot\phi^2 + \tfrac{1}{2}(I_1 \cos^2\alpha + I_2 \cos^2\beta + I_3 \cos^2\gamma)\dot\phi^2 - \tfrac{1}{2}\mu g l \phi^2.$$

The frequency of the oscillations is consequently

$$\omega^2 = \mu g l/(\mu l^2 + I_1 \cos^2\alpha + I_2 \cos^2\beta + I_3 \cos^2\gamma).$$

PROBLEM 4. Find the kinetic energy of the system shown in Fig. 39: OA and AB are thin uniform rods of length l hinged together at A. The rod OA rotates (in the plane of the diagram) about O, while the end B of the rod AB slides along Ox.

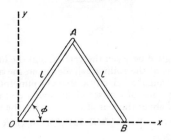

FIG. 39

SOLUTION. The velocity of the centre of mass of the rod OA (which is at the middle of the rod) is $\frac{1}{2}l\dot\phi$, where ϕ is the angle AOB. The kinetic energy of the rod OA is therefore $T_1 = \frac{1}{8}\mu l^2\dot\phi^2 + \frac{1}{2}I\dot\phi^2$, where μ is the mass of each rod.

The Cartesian co-ordinates of the centre of mass of the rod AB are $X = \frac{3}{2}l \cos \phi$, $Y = \frac{1}{2}l \sin \phi$. Since the angular velocity of rotation of this rod is also $\dot\phi$, its kinetic energy is $T_2 = \frac{1}{2}\mu(\dot{X}^2 + \dot{Y}^2) + \frac{1}{2}I\dot\phi^2 = \frac{1}{8}\mu l^2(1 + 8 \sin^2\phi)\dot\phi^2 + \frac{1}{2}I\dot\phi^2$. The total kinetic energy of this system is therefore $T = \frac{1}{6}\mu l^2(1 + 3 \sin^2\phi)\dot\phi^2$, since $I = \frac{1}{12}\mu l^2$ (see Problem 2(a)).

PROBLEM 5. Find the kinetic energy of a cylinder of radius R rolling on a plane, if the mass of the cylinder is so distributed that one of the principal axes of inertia is parallel to the axis of the cylinder and at a distance a from it, and the moment of inertia about that principal axis is I.

SOLUTION. Let ϕ be the angle between the vertical and a line from the centre of mass perpendicular to the axis of the cylinder (Fig. 40). The motion of the cylinder at any instant

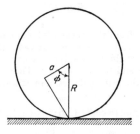

FIG. 40

may be regarded as a pure rotation about an instantaneous axis which coincides with the line where the cylinder touches the plane. The angular velocity of this rotation is $\dot\phi$, since the angular velocity of rotation about all parallel axes is the same. The centre of mass is at a distance $\sqrt{(a^2 + R^2 - 2aR \cos \phi)}$ from the instantaneous axis, and its velocity is therefore $V = \dot\phi\sqrt{(a^2 + R^2 - 2aR \cos \phi)}$. The total kinetic energy is

$$T = \tfrac{1}{2}\mu(a^2 + R^2 - 2aR \cos \phi)\dot\phi^2 + \tfrac{1}{2}I\dot\phi^2.$$

PROBLEM 6. Find the kinetic energy of a homogeneous cylinder of radius a rolling inside a cylindrical surface of radius R (Fig. 41).

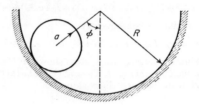

FIG. 41

SOLUTION. We use the angle ϕ between the vertical and the line joining the centres of the cylinders. The centre of mass of the rolling cylinder is on the axis, and its velocity is $V = \dot\phi(R-a)$. We can calculate the angular velocity as that of a pure rotation about an instantaneous axis which coincides with the line of contact of the cylinders; it is $\Omega = V/a = \dot\phi(R-a)/a$. If I_3 is the moment of inertia about the axis of the cylinder, then

$$T = \tfrac{1}{2}\mu(R-a)^2\dot\phi^2 + \tfrac{1}{2}I_3(R-a)^2\dot\phi^2/a^2 = \tfrac{3}{4}\mu(R-a)^2\dot\phi^2,$$

I_3 being given by Problem 2(c).

PROBLEM 7. Find the kinetic energy of a homogeneous cone rolling on a plane.

SOLUTION. We denote by θ the angle between the line OA in which the cone touches the plane and some fixed direction in the plane (Fig. 42). The centre of mass is on the axis of the cone, and its velocity $V = a\dot\theta\cos\alpha$, where 2α is the vertical angle of the cone and a the

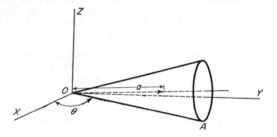

FIG. 42

distance of the centre of mass from the vertex. The angular velocity can be calculated as that of a pure rotation about the instantaneous axis OA: $\Omega = V/a\sin\alpha = \dot\theta\cot\alpha$. One of the principal axes of inertia (x_3) is along the axis of the cone, and we take another (x_2) perpendicular to the axis of the cone and to the line OA. Then the components of the vector $\boldsymbol\Omega$ (which is parallel to OA) along the principal axes of inertia are $\Omega\sin\alpha$, 0, $\Omega\cos\alpha$. The kinetic energy is thus

$$T = \tfrac{1}{2}\mu a^2\dot\theta^2\cos^2\alpha + \tfrac{1}{2}I_1\dot\theta^2\cos^2\alpha + \tfrac{1}{2}I_3\dot\theta^2\frac{\cos^4\alpha}{\sin^2\alpha}$$

$$= 3\mu h^2\dot\theta^2(1+5\cos^2\alpha)/40,$$

where h is the height of the cone, and I_1, I_3 and a have been given in Problem 2(e).

PROBLEM 8. Find the kinetic energy of a homogeneous cone whose base rolls on a plane and whose vertex is fixed at a height above the plane equal to the radius of the base, so that the axis of the cone is parallel to the plane.

SOLUTION. We use the angle θ between a fixed direction in the plane and the projection of the axis of the cone on the plane (Fig. 43). Then the velocity of the centre of mass is $V = a\dot\theta$, the notation being as in Problem 7. The instantaneous axis of rotation is the generator OA which passes through the point where the cone touches the plane. The centre of mass is at a distance $a \sin \alpha$ from this axis, and so $\Omega = V/a \sin \alpha = \dot\theta/\sin \alpha$. The components of the vector $\boldsymbol\Omega$ along the principal axes of inertia are, if the x_2-axis is taken perpendicular to the axis of the cone and to the line OA, $\Omega \sin \alpha = \dot\theta$, 0, $\Omega \cos \alpha = \dot\theta \cot \alpha$. The kinetic energy is therefore

$$T = \tfrac{1}{2}\mu a^2\dot\theta^2 + \tfrac{1}{2}I_1\dot\theta^2 + \tfrac{1}{2}I_3\dot\theta^2 \cot^2\alpha$$
$$= 3\mu h^2\dot\theta^2(\sec^2\alpha + 5)/40.$$

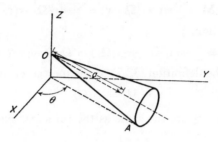

FIG. 43

PROBLEM 9. Find the kinetic energy of a homogeneous ellipsoid which rotates about one of its axes (AB in Fig. 44) while that axis itself rotates about a line CD perpendicular to it and passing through the centre of the ellipsoid.

SOLUTION. Let the angle of rotation about CD be θ, and that about AB (i.e. the angle between CD and the x_1-axis of inertia, which is perpendicular to AB) be ϕ. Then the components of $\boldsymbol\Omega$ along the axes of inertia are $\dot\theta \cos \phi$, $\dot\theta \sin \phi$, $\dot\phi$, if the x_3-axis is AB. Since the centre of mass, at the centre of the ellipsoid, is at rest, the kinetic energy is

$$T = \tfrac{1}{2}(I_1 \cos^2\phi + I_2 \sin^2\phi)\dot\theta^2 + \tfrac{1}{2}I_3\dot\phi^2.$$

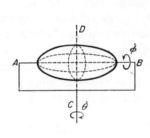   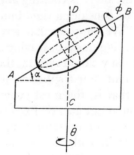

FIG. 44   FIG. 45

PROBLEM 10. The same as Problem 9, but for the case where the axis AB is not perpendicular to CD and is an axis of symmetry of the ellipsoid (Fig. 45).

SOLUTION. The components of $\boldsymbol\Omega$ along the axis AB and the other two principal axes of inertia, which are perpendicular to AB but otherwise arbitrary, are $\dot\theta \cos \alpha \cos \phi$, $\dot\theta \cos \alpha \times \sin \phi$, $\dot\phi + \dot\theta \sin \alpha$. The kinetic energy is $T = \tfrac{1}{2}I_1\dot\theta^2 \cos^2\alpha + \tfrac{1}{2}I_3(\dot\phi + \dot\theta \sin \alpha)^2$.

§33. Angular momentum of a rigid body

The value of the angular momentum of a system depends, as we know, on the point with respect to which it is defined. In the mechanics of a rigid body, the most appropriate point to choose for this purpose is the origin of the moving system of co-ordinates, i.e. the centre of mass of the body, and in what follows we shall denote by **M** the angular momentum so defined.

According to formula (9.6), when the origin is taken at the centre of mass of the body, the angular momentum **M** is equal to the "intrinsic" angular momentum resulting from the motion relative to the centre of mass. In the definition $\mathbf{M} = \Sigma m\mathbf{r} \times \mathbf{v}$ we therefore replace **v** by $\boldsymbol{\Omega} \times \mathbf{r}$:

$$\mathbf{M} = \sum m\mathbf{r} \times (\boldsymbol{\Omega} \times \mathbf{r}) = \sum m[r^2 \boldsymbol{\Omega} - \mathbf{r}(\mathbf{r} \cdot \boldsymbol{\Omega})],$$

or, in tensor notation,

$$M_i = \sum m(x_l^2 \Omega_i - x_i x_k \Omega_k) = \Omega_k \sum m(x_l^2 \delta_{ik} - x_i x_k).$$

Finally, using the definition (32.2) of the inertia tensor, we have

$$M_i = I_{ik} \Omega_k. \tag{33.1}$$

If the axes x_1, x_2, x_3 are the same as the principal axes of inertia, formula (33.1) gives

$$M_1 = I_1 \Omega_1, \qquad M_2 = I_2 \Omega_2, \qquad M_3 = I_3 \Omega_3. \tag{33.2}$$

In particular, for a spherical top, where all three principal moments of inertia are equal, we have simply

$$\mathbf{M} = I\boldsymbol{\Omega}, \tag{33.3}$$

i.e. the angular momentum vector is proportional to, and in the same direction as, the angular velocity vector. For an arbitrary body, however, the vector **M** is not in general in the same direction as $\boldsymbol{\Omega}$; this happens only when the body is rotating about one of its principal axes of inertia.

Let us consider a rigid body moving freely, i.e. not subject to any external forces. We suppose that any uniform translational motion, which is of no interest, is removed, leaving a free rotation of the body.

As in any closed system, the angular momentum of the freely rotating body is constant. For a spherical top the condition **M** = constant gives $\boldsymbol{\Omega}$ = constant; that is, the most general free rotation of a spherical top is a uniform rotation about an axis fixed in space.

The case of a rotator is equally simple. Here also $\mathbf{M} = I\boldsymbol{\Omega}$, and the vector $\boldsymbol{\Omega}$ is perpendicular to the axis of the rotator. Hence a free rotation of a rotator is a uniform rotation in one plane about an axis perpendicular to that plane.

The law of conservation of angular momentum also suffices to determine the more complex free rotation of a symmetrical top. Using the fact that the principal axes of inertia x_1, x_2 (perpendicular to the axis of symmetry (x_3) of the top) may be chosen arbitrarily, we take the x_2-axis perpendicular to the plane containing the constant vector **M** and the instantaneous position of the x_3-axis. Then $M_2 = 0$, and formulae (33.2) show that $\Omega_2 = 0$. This means that the directions of **M**, $\boldsymbol{\Omega}$ and the axis of the top are at every instant in one plane (Fig. 46). Hence, in turn, it follows that the velocity $\mathbf{v} = \boldsymbol{\Omega} \times \mathbf{r}$ of every point on the axis of the top is at every instant perpendicular to that plane. That is, the axis of the top rotates uniformly (see below) about the direction of **M**, describing a circular cone. This is called *regular precession* of the top. At the same time the top rotates uniformly about its own axis.

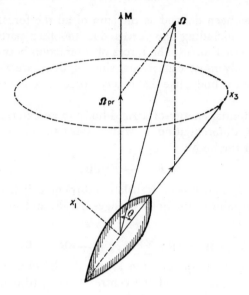

Fɪɢ. 46

The angular velocities of these two rotations can easily be expressed in terms of the given angular momentum **M** and the angle θ between the axis of the top and the direction of **M**. The angular velocity of the top about its own axis is just the component Ω_3 of the vector $\boldsymbol{\Omega}$ along the axis:

$$\Omega_3 = M_3/I_3 = (M/I_3) \cos \theta. \tag{33.4}$$

To determine the rate of precession Ω_{pr}, the vector $\boldsymbol{\Omega}$ must be resolved into components along x_3 and along **M**. The first of these gives no displacement of the axis of the top, and the second component is therefore the required angular velocity of precession. Fig. 46 shows that $\Omega_{pr} \sin \theta = \Omega_1$, and, since $\Omega_1 = M_1/I_1 = (M/I_1) \sin \theta$, we have

$$\Omega_{pr} = M/I_1. \tag{33.5}$$

§34. The equations of motion of a rigid body

Since a rigid body has, in general, six degrees of freedom, the general equations of motion must be six in number. They can be put in a form which gives the time derivatives of two vectors, the momentum and the angular momentum of the body.

The first equation is obtained by simply summing the equations $\dot{\mathbf{p}} = \mathbf{f}$ for each particle in the body, **p** being the momentum of the particle and **f** the force acting on it. In terms of the total momentum of the body $\mathbf{P} = \Sigma\mathbf{p} = \mu\mathbf{V}$ and total force acting on it $\mathbf{F} = \Sigma\mathbf{f}$, we have

$$d\mathbf{P}/dt = \mathbf{F}. \tag{34.1}$$

Although $\mathbf{F}$ has been defined as the sum of all the forces $\mathbf{f}$ acting on the various particles, including the forces due to other particles, $\mathbf{F}$ actually includes only external forces: the forces of interaction between the particles composing the body must cancel out, since if there are no external forces the momentum of the body, like that of any closed system, must be conserved, i.e. we must have $\mathbf{F} = 0$.

If U is the potential energy of a rigid body in an external field, the force $\mathbf{F}$ is obtained by differentiating U with respect to the co-ordinates of the centre of mass of the body:

$$\mathbf{F} = -\partial U/\partial \mathbf{R}. \tag{34.2}$$

For, when the body undergoes a translation through a distance $\delta\mathbf{R}$, the radius vector $\mathbf{r}$ of every point in the body changes by $\delta\mathbf{R}$, and so the change in the potential energy is

$$\delta U = \sum(\partial U/\partial \mathbf{r}) \cdot \delta\mathbf{r} = \delta\mathbf{R} \cdot \sum \partial U/\partial \mathbf{r} = -\delta\mathbf{R} \cdot \sum \mathbf{f} = -\mathbf{F} \cdot \delta\mathbf{R}.$$

It may be noted that equation (34.1) can also be obtained as Lagrange's equation for the co-ordinates of the centre of mass, $(d/dt)\partial L/\partial \mathbf{V} = \partial L/\partial \mathbf{R}$, with the Lagrangian (32.4), for which

$$\partial L/\partial \mathbf{V} = \mu\mathbf{V} = \mathbf{P}, \qquad \partial L/\partial \mathbf{R} = -\partial U/\partial \mathbf{R} = \mathbf{F}.$$

Let us now derive the second equation of motion, which gives the time derivative of the angular momentum $\mathbf{M}$. To simplify the derivation, it is convenient to choose the "fixed" (inertial) frame of reference in such a way that the centre of mass is at rest in that frame at the instant considered.

We have $\dot{\mathbf{M}} = (d/dt)\Sigma\mathbf{r}\times\mathbf{p} = \Sigma\dot{\mathbf{r}}\times\mathbf{p}+\Sigma\mathbf{r}\times\dot{\mathbf{p}}$. Our choice of the frame of reference (with $\mathbf{V} = 0$) means that the value of $\dot{\mathbf{r}}$ at the instant considered is the same as $\mathbf{v} = \dot{\mathbf{r}}$. Since the vectors $\mathbf{v}$ and $\mathbf{p} = m\mathbf{v}$ are parallel, $\dot{\mathbf{r}}\times\mathbf{p} = 0$. Replacing $\dot{\mathbf{p}}$ by the force $\mathbf{f}$, we have finally

$$d\mathbf{M}/dt = \mathbf{K}, \tag{34.3}$$

where

$$\mathbf{K} = \sum \mathbf{r}\times\mathbf{f}. \tag{34.4}$$

Since $\mathbf{M}$ has been defined as the angular momentum about the centre of mass (see the beginning of §33), it is unchanged when we go from one inertial frame to another. This is seen from formula (9.5) with $\mathbf{R} = 0$. We can therefore deduce that the equation of motion (34.3), though derived for a particular frame of reference, is valid in any other inertial frame, by Galileo's relativity principle.

The vector $\mathbf{r}\times\mathbf{f}$ is called the *moment* of the force $\mathbf{f}$, and so $\mathbf{K}$ is the total *torque*, i.e. the sum of the moments of all the forces acting on the body. Like the total force **F**, the sum (34.4) need include only the external forces: by the law of conservation of angular momentum, the sum of the moments of the internal forces in a closed system must be zero.

The moment of a force, like the angular momentum, in general depends on the choice of the origin about which it is defined. In (34.3) and (34.4) the moments are defined with respect to the centre of mass of the body.

When the origin is moved a distance **a**, the new radius vector **r**' of each point in the body is equal to **r** − **a**. Hence $\mathbf{K} = \Sigma \mathbf{r} \times \mathbf{f} = \Sigma \mathbf{r}' \times \mathbf{f} + \Sigma \mathbf{a} \times \mathbf{f}$ or

$$\mathbf{K} = \mathbf{K}' + \mathbf{a} \times \mathbf{F}. \tag{34.5}$$

Hence we see, in particular, that the value of the torque is independent of the choice of origin if the total force **F** = 0. In this case the body is said to be acted on by a *couple*.

Equation (34.3) may be regarded as Lagrange's equation $(d/dt) \, \partial L/\partial \mathbf{\Omega} = \partial L/\partial \boldsymbol{\phi}$ for the "rotational co-ordinates". Differentiating the Lagrangian (32.4) with respect to the components of the vector **Ω**, we obtain $\partial L/\partial \Omega_i = I_{ik}\Omega_k = M_i$. The change in the potential energy resulting from an infinitesimal rotation $\delta\boldsymbol{\phi}$ of the body is $\delta U = -\Sigma \mathbf{f} \cdot \delta \mathbf{r} = -\Sigma \mathbf{f} \cdot \delta\boldsymbol{\phi} \times \mathbf{r} = -\delta\boldsymbol{\phi} \cdot \Sigma \mathbf{r} \times \mathbf{f} = -\mathbf{K} \cdot \delta\boldsymbol{\phi}$, whence

$$\mathbf{K} = -\partial U/\partial\boldsymbol{\phi}, \tag{34.6}$$

so that $\partial L/\partial\boldsymbol{\phi} = -\partial U/\partial\boldsymbol{\phi} = \mathbf{K}$.

Let us assume that the vectors **F** and **K** are perpendicular. Then a vector **a** can always be found such that **K**' given by formula (34.5) is zero and

$$\mathbf{K} = \mathbf{a} \times \mathbf{F}. \tag{34.7}$$

The choice of **a** is not unique, since the addition to **a** of any vector parallel to **F** does not affect equation (34.7). The condition **K**' = 0 thus gives a straight line, not a point, in the moving system of co-ordinates. When **K** is perpendicular to **F**, the effect of all the applied forces can therefore be reduced to that of a single force **F** acting along this line.

Such a case is that of a uniform field of force, in which the force on a particle is **f** = *e***E**, with **E** a constant vector characterising the field and *e* characterising the properties of a particle with respect to the field.† Then **F** = **E**Σ*e*, **K** = Σ*e***r** × **E**. Assuming that Σ*e* ≠ 0, we define a radius vector **r**₀ such that

$$\mathbf{r}_0 = \sum er \Big/ \sum e. \tag{34.8}$$

Then the total torque is simply

$$\mathbf{K} = \mathbf{r}_0 \times \mathbf{F}. \tag{34.9}$$

Thus, when a rigid body moves in a uniform field, the effect of the field reduces to the action of a single force **F** applied at the point whose radius vector is (34.8). The position of this point is entirely determined by the

---

† For example, in a uniform electric field **E** is the field strength and *e* the charge; in a uniform gravitational field **E** is the acceleration **g** due to gravity and *e* is the mass *m*.

properties of the body itself. In a gravitational field, for example, it is the centre of mass.

§35. Eulerian angles

As has already been mentioned, the motion of a rigid body can be described by means of the three co-ordinates of its centre of mass and any three angles which determine the orientation of the axes x_1, x_2, x_3 in the moving system of co-ordinates relative to the fixed system X, Y, Z. These angles may often be conveniently taken as what are called *Eulerian angles*.

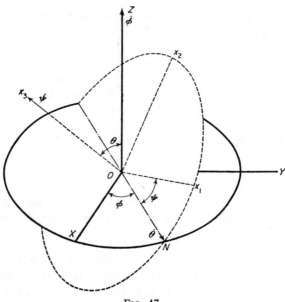

FIG. 47

Since we are here interested only in the angles between the co-ordinate axes, we may take the origins of the two systems to coincide (Fig. 47). The moving x_1x_2-plane intersects the fixed XY-plane in some line ON, called the *line of nodes*. This line is evidently perpendicular to both the Z-axis and the x_3-axis; we take its positive direction as that of the vector product $\mathbf{z} \times \mathbf{x_3}$ (where $\mathbf{z}$ and $\mathbf{x_3}$ are unit vectors along the Z and x_3 axes).

We take, as the quantities defining the position of the axes x_1, x_2, x_3 relative to the axes X, Y, Z the angle θ between the Z and x_3 axes, the angle ϕ between the X-axis and ON, and the angle ψ between the x_1-axis and ON. The angles ϕ and ψ are measured round the Z and x_3 axes respectively in the direction given by the corkscrew rule. The angle θ takes values from 0 to π, and ϕ and ψ from 0 to 2π.†

---

† The angles θ and $\phi - \frac{1}{2}\pi$ are respectively the polar angle and azimuth of the direction x_3 with respect to the axes X, Y, Z. The angles θ and $\frac{1}{2}\pi - \psi$ are respectively the polar angle and azimuth of the direction Z with respect to the axes x_1, x_2, x_3.

Let us now express the components of the angular velocity vector $\mathbf{\Omega}$ along the moving axes x_1, x_2, x_3 in terms of the Eulerian angles and their derivatives. To do this, we must find the components along those axes of the angular velocities $\dot{\theta}$, $\dot{\phi}$, $\dot{\psi}$. The angular velocity $\dot{\theta}$ is along the line of nodes ON, and its components are $\dot{\theta}_1 = \dot{\theta} \cos \psi$, $\dot{\theta}_2 = -\dot{\theta} \sin \psi$, $\dot{\theta}_3 = 0$. The angular velocity $\dot{\phi}$ is along the Z-axis; its component along the x_3-axis is $\dot{\phi}_3 = \dot{\phi} \cos \theta$, and in the $x_1 x_2$-plane $\dot{\phi} \sin \theta$. Resolving the latter along the x_1 and x_2 axes, we have $\dot{\phi}_1 = \dot{\phi} \sin \theta \sin \psi$, $\dot{\phi}_2 = \dot{\phi} \sin \theta \cos \psi$. Finally, the angular velocity $\dot{\psi}$ is along the x_3-axis.

Collecting the components along each axis, we have

$$\left. \begin{aligned} \Omega_1 &= \dot{\phi} \sin \theta \sin \psi + \dot{\theta} \cos \psi, \\ \Omega_2 &= \dot{\phi} \sin \theta \cos \psi - \dot{\theta} \sin \psi, \\ \Omega_3 &= \dot{\phi} \cos \theta + \dot{\psi}. \end{aligned} \right\} \tag{35.1}$$

If the axes x_1, x_2, x_3 are taken to be the principal axes of inertia of the body, the rotational kinetic energy in terms of the Eulerian angles is obtained by substituting (35.1) in (32.8).

For a symmetrical top ($I_1 = I_2 \neq I_3$), a simple reduction gives

$$T_{\text{rot}} = \tfrac{1}{2} I_1 (\dot{\phi}^2 \sin^2 \theta + \dot{\theta}^2) + \tfrac{1}{2} I_3 (\dot{\phi} \cos \theta + \dot{\psi})^2. \tag{35.2}$$

This expression can also be more simply obtained by using the fact that the choice of directions of the principal axes x_1, x_2 is arbitrary for a symmetrical top. If the x_1 axis is taken along the line of nodes ON, i.e. $\psi = 0$, the components of the angular velocity are simply

$$\Omega_1 = \dot{\theta}, \qquad \Omega_2 = \dot{\phi} \sin \theta, \qquad \Omega_3 = \dot{\phi} \cos \theta + \dot{\psi}. \tag{35.3}$$

As a simple example of the use of the Eulerian angles, we shall use them to determine the free motion of a symmetrical top, already found in §33. We take the Z-axis of the fixed system of co-ordinates in the direction of the constant angular momentum $\mathbf{M}$ of the top. The x_3-axis of the moving system is along the axis of the top; let the x_1-axis coincide with the line of nodes at the instant considered. Then the components of the vector $\mathbf{M}$ are, by formulae (35.3), $M_1 = I_1 \Omega_1 = I_1 \dot{\theta}$, $M_2 = I_1 \Omega_2 = I_1 \dot{\phi} \sin \theta$, $M_3 = I_3 \Omega_3 = I_3 (\dot{\phi} \cos \theta + \dot{\psi})$. Since the x_1-axis is perpendicular to the Z-axis, we have $M_1 = 0$, $M_2 = M \sin \theta$, $M_3 = M \cos \theta$. Comparison gives

$$\dot{\theta} = 0, \qquad I_1 \dot{\phi} = M, \qquad I_3 (\dot{\phi} \cos \theta + \dot{\psi}) = M \cos \theta. \tag{35.4}$$

The first of these equations gives $\theta = \text{constant}$, i.e. the angle between the axis of the top and the direction of $\mathbf{M}$ is constant. The second equation gives the angular velocity of precession $\dot{\phi} = M/I_1$, in agreement with (33.5). Finally, the third equation gives the angular velocity with which the top rotates about its own axis: $\Omega_3 = (M/I_3) \cos \theta$.

PROBLEMS

PROBLEM 1. Reduce to quadratures the problem of the motion of a heavy symmetrical top whose lowest point is fixed (Fig. 48).

SOLUTION. We take the common origin of the moving and fixed systems of co-ordinates at the fixed point O of the top, and the Z-axis vertical. The Lagrangian of the top in a gravitational field is $L = \frac{1}{2}(I_1 + \mu l^2)(\dot{\theta}^2 + \dot{\phi}^2 \sin^2\theta) + \frac{1}{2}I_3(\dot{\psi} + \dot{\phi} \cos \theta)^2 - \mu gl \cos \theta$, where μ is the mass of the top and l the distance from its fixed point to the centre of mass.

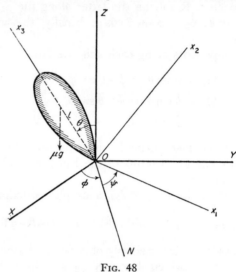

FIG. 48

The co-ordinates ψ and ϕ are cyclic. Hence we have two integrals of the motion:

$$p_\psi = \partial L/\partial\dot{\psi} = I_3(\dot{\psi} + \dot{\phi} \cos \theta) = \text{constant} \equiv M_3 \tag{1}$$

$$p_\phi = \partial L/\partial\dot{\phi} = (I'_1 \sin^2\theta + I_3 \cos^2\theta)\dot{\phi} + I_3\dot{\psi} \cos \theta = \text{constant} \equiv M_z, \tag{2}$$

where $I'_1 = I_1 + \mu l^2$; the quantities p_ψ and p_ϕ are the components of the rotational angular momentum about O along the x_3 and Z axes respectively. The energy

$$E = \frac{1}{2}I'_1(\dot{\theta}^2 + \dot{\phi}^2 \sin^2\theta) + \frac{1}{2}I_3(\dot{\psi} + \dot{\phi} \cos \theta)^2 + \mu gl \cos \theta \tag{3}$$

is also conserved.

From equations (1) and (2) we find

$$\dot{\phi} = (M_z - M_3 \cos \theta)/I'_1 \sin^2\theta, \tag{4}$$

$$\dot{\psi} = \frac{M_3}{I_3} - \cos \theta \frac{M_z - M_3 \cos \theta}{I'_1 \sin^2\theta}. \tag{5}$$

Eliminating $\dot{\phi}$ and $\dot{\psi}$ from the energy (3) by means of equations (4) and (5), we obtain

$$E' = \frac{1}{2}I'_1\dot{\theta}^2 + U_{\text{eff}}(\theta),$$

where

$$E' = E - \frac{M_3^2}{2I_3} - \mu gl,$$

$$U_{\text{eff}} = \frac{(M_z - M_3 \cos \theta)^2}{2I'_1 \sin^2\theta} - \mu gl(1 - \cos \theta). \tag{6}$$

Thus we have

$$t = \int \frac{d\theta}{\sqrt{\{2[E' - U_{\text{eff}}(\theta)]/I'_1\}}};$$

(7)

this is an elliptic integral. The angles ψ and ϕ are then expressed in terms of θ by means of integrals obtained from equations (4) and (5).

The range of variation of θ during the motion is determined by the condition $E' \geqslant U_{\text{eff}}(\theta)$. The function $U_{\text{eff}}(\theta)$ tends to infinity (if $M_3 \neq M_z$) when θ tends to 0 or π, and has a minimum between these values. Hence the equation $E' = U_{\text{eff}}(\theta)$ has two roots, which determine the limiting values θ_1 and θ_2 of the inclination of the axis of the top to the vertical.

When θ varies from θ_1 to θ_2, the derivative $\dot\phi$ changes sign if and only if the difference $M_z - M_3 \cos\theta$ changes sign in that range of θ. If it does not change sign, the axis of the top precesses monotonically about the vertical, at the same time oscillating up and down. The latter oscillation is called *nutation*; see Fig. 49a, where the curve shows the track of the axis on the surface of a sphere whose centre is at the fixed point of the top. If $\dot\phi$ does change sign, the direction of precession is opposite on the two limiting circles, and so the axis of the top describes loops as it moves round the vertical (Fig. 49b). Finally, if one of θ_1, θ_2 is a zero of $M_z - M_3 \cos\theta$, $\dot\phi$ and $\dot\theta$ vanish together on the corresponding limiting circle, and the path of the axis is of the kind shown in Fig. 49c.

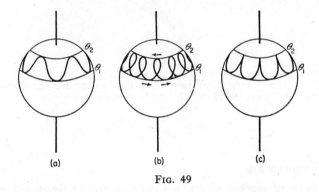

(a)   (b)   (c)

FIG. 49

PROBLEM 2. Find the condition for the rotation of a top about a vertical axis to be stable.

SOLUTION. For $\theta = 0$, the x_3 and Z axes coincide, so that $M_3 = M_z$, $E' = 0$. Rotation about this axis is stable if $\theta = 0$ is a minimum of the function $U_{\text{eff}}(\theta)$. For small θ we have $U_{\text{eff}} \approx (M_3^2/8I'_1 - \tfrac{1}{2}\mu gl)\theta^2$, whence the condition for stability is $M_3^2 > 4I'_1\mu gl$ or $\Omega_3^2 > 4I'_1\mu gl/I_3^2$.

PROBLEM 3. Determine the motion of a top when the kinetic energy of its rotation about its axis is large compared with its energy in the gravitational field (called a *"fast"* top).

SOLUTION. In a first approximation, neglecting gravity, there is a free precession of the axis of the top about the direction of the angular momentum $\mathbf{M}$, corresponding in this case to the nutation of the top; according to (33.5), the angular velocity of this precession is

$$\Omega_{\text{nu}} = M/I'_1.$$

(1)

In the next approximation, there is a slow precession of the angular momentum $\mathbf{M}$ about the vertical (Fig. 50). To determine the rate of this precession, we average the exact equation of motion (34.3) $d\mathbf{M}/dt = \mathbf{K}$ over the nutation period. The moment of the force of gravity on the top is $\mathbf{K} = \mu l \mathbf{n}_3 \times \mathbf{g}$, where $\mathbf{n}_3$ is a unit vector along the axis of the top. It is evident from symmetry that the result of averaging $\mathbf{K}$ over the "nutation cone" is to replace $\mathbf{n}_3$ by its component $(\mathbf{M}/M)\cos\alpha$ in the direction of $\mathbf{M}$, where α is the angle between $\mathbf{M}$ and the axis of the top. Thus we have $\overline{d\mathbf{M}/dt} = -(\mu l/M)\mathbf{g} \times \mathbf{M} \cos\alpha$. This shows that the vector $\mathbf{M}$

precesses about the direction of **g** (i.e. the vertical) with a mean angular velocity

$$\bar{\boldsymbol{\Omega}}_{\text{pr}} = -(\mu l/M)\mathbf{g}\cos\alpha \qquad (2)$$

which is small compared with $\boldsymbol{\Omega}_{\text{nu}}$.

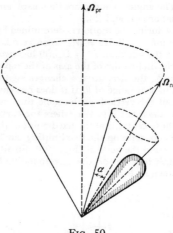

Fig. 50

In this approximation the quantities M and $\cos\alpha$ in formulae (1) and (2) are constants, although they are not exact integrals of the motion. To the same accuracy they are related to the strictly conserved quantities E and M_3 by $M_3 = M\cos\alpha$,

$$E \approx \tfrac{1}{2}M^2\left(\frac{\cos^2\alpha}{I_3} + \frac{\sin^2\alpha}{I'_1}\right).$$

§36. Euler's equations

The equations of motion given in §34 relate to the fixed system of co-ordinates: the derivatives $d\mathbf{P}/dt$ and $d\mathbf{M}/dt$ in equations (34.1) and (34.3) are the rates of change of the vectors $\mathbf{P}$ and $\mathbf{M}$ with respect to that system. The simplest relation between the components of the rotational angular momentum $\mathbf{M}$ of a rigid body and the components of the angular velocity occurs, however, in the moving system of co-ordinates whose axes are the principal axes of inertia. In order to use this relation, we must first transform the equations of motion to the moving co-ordinates x_1, x_2, x_3.

Let $d\mathbf{A}/dt$ be the rate of change of any vector $\mathbf{A}$ with respect to the fixed system of co-ordinates. If the vector $\mathbf{A}$ does not change in the moving system, its rate of change in the fixed system is due only to the rotation, so that $d\mathbf{A}/dt = \boldsymbol{\Omega}\times\mathbf{A}$; see §9, where it has been pointed out that formulae such as (9.1) and (9.2) are valid for any vector. In the general case, the right-hand side includes also the rate of change of the vector $\mathbf{A}$ with respect to the moving system. Denoting this rate of change by $d'\mathbf{A}/dt$, we obtain

$$\frac{d\mathbf{A}}{dt} = \frac{d'\mathbf{A}}{dt} + \boldsymbol{\Omega}\times\mathbf{A}. \qquad (36.1)$$

Using this general formula, we can immediately write equations (34.1) and (34.3) in the form

$$\frac{d'\mathbf{P}}{dt}+\mathbf{\Omega}\times\mathbf{P} = \mathbf{F}, \qquad \frac{d'\mathbf{M}}{dt}+\mathbf{\Omega}\times\mathbf{M} = \mathbf{K}. \tag{36.2}$$

Since the differentiation with respect to time is here performed in the moving system of co-ordinates, we can take the components of equations (36.2) along the axes of that system, putting $(d'\mathbf{P}/dt)_1 = dP_1/dt, \ldots, (d'\mathbf{M}/dt)_1 = dM_1/dt,$ ..., where the suffixes 1, 2, 3 denote the components along the axes x_1, x_2, x_3. In the first equation we replace $\mathbf{P}$ by $\mu\mathbf{V}$, obtaining

$$\left.\begin{aligned}
\mu\left(\frac{dV_1}{dt}+\Omega_2 V_3-\Omega_3 V_2\right) &= F_1, \\[2mm]
\mu\left(\frac{dV_2}{dt}+\Omega_3 V_1-\Omega_1 V_3\right) &= F_2, \\[2mm]
\mu\left(\frac{dV_3}{dt}+\Omega_1 V_2-\Omega_2 V_1\right) &= F_3.
\end{aligned}\right\} \tag{36.3}$$

If the axes x_1, x_2, x_3 are the principal axes of inertia, we can put $M_1 = I_1\Omega_1$, etc., in the second equation (36.2), obtaining

$$\left.\begin{aligned}
I_1\, d\Omega_1/dt+(I_3-I_2)\Omega_2\Omega_3 &= K_1, \\
I_2\, d\Omega_2/dt+(I_1-I_3)\Omega_3\Omega_1 &= K_2, \\
I_3\, d\Omega_3/dt+(I_2-I_1)\Omega_1\Omega_2 &= K_3.
\end{aligned}\right\} \tag{36.4}$$

These are *Euler's equations.*

In free rotation, $\mathbf{K} = 0$, so that Euler's equations become

$$\left.\begin{aligned}
d\Omega_1/dt+(I_3-I_2)\Omega_2\Omega_3/I_1 &= 0, \\
d\Omega_2/dt+(I_1-I_3)\Omega_3\Omega_1/I_2 &= 0, \\
d\Omega_3/dt+(I_2-I_1)\Omega_1\Omega_2/I_3 &= 0.
\end{aligned}\right\} \tag{36.5}$$

As an example, let us apply these equations to the free rotation of a symmetrical top, which has already been discussed. Putting $I_1 = I_2$, we find from the third equation $\dot{\Omega}_3 = 0$, i.e. $\Omega_3 = $ constant. We then write the first two equations as $\dot{\Omega}_1 = -\omega\Omega_2$, $\dot{\Omega}_2 = \omega\Omega_1$, where

$$\omega = \Omega_3(I_3-I_1)/I_1 \tag{36.6}$$

is a constant. Multiplying the second equation by i and adding, we have $d(\Omega_1+i\Omega_2)/dt = i\omega(\Omega_1+i\Omega_2)$, so that $\Omega_1+i\Omega_2 = A\exp(i\omega t)$, where A is a constant, which may be made real by a suitable choice of the origin of time. Thus

$$\Omega_1 = A\cos\omega t \qquad \Omega_2 = A\sin\omega t. \tag{36.7}$$

This result shows that the component of the angular velocity perpendicular to the axis of the top rotates with an angular velocity ω, remaining of constant magnitude $A = \sqrt{(\Omega_1^2 + \Omega_2^2)}$. Since the component Ω_3 along the axis of the top is also constant, we conclude that the vector $\boldsymbol{\Omega}$ rotates uniformly with angular velocity ω about the axis of the top, remaining unchanged in magnitude. On account of the relations $M_1 = I_1\Omega_1$, $M_2 = I_2\Omega_2$, $M_3 = I_3\Omega_3$ between the components of $\boldsymbol{\Omega}$ and $\mathbf{M}$, the angular momentum vector $\mathbf{M}$ evidently executes a similar motion with respect to the axis of the top.

This description is naturally only a different view of the motion already discussed in §33 and §35, where it was referred to the fixed system of co-ordinates. In particular, the angular velocity of the vector $\mathbf{M}$ (the Z-axis in Fig. 48, §35) about the x_3-axis is, in terms of Eulerian angles, the same as the angular velocity $-\dot{\psi}$. Using equations (35.4), we have

$$\dot{\psi} = \frac{M\cos\theta}{I_3} - \dot{\phi}\cos\theta = M\cos\theta\left(\frac{1}{I_3} - \frac{1}{I_1}\right),$$

or $-\dot{\psi} = \Omega_3(I_3 - I_1)/I_1$, in agreement with (36.6).

§37. The asymmetrical top

We shall now apply Euler's equations to the still more complex problem of the free rotation of an asymmetrical top, for which all three moments of inertia are different. We assume for definiteness that

$$I_3 > I_2 > I_1. \tag{37.1}$$

Two integrals of Euler's equations are known already from the laws of conservation of energy and angular momentum:

$$I_1\Omega_1^2 + I_2\Omega_2^2 + I_3\Omega_3^2 = 2E,$$
$$I_1^2\Omega_1^2 + I_2^2\Omega_2^2 + I_3^2\Omega_3^2 = M^2, \tag{37.2}$$

where the energy E and the magnitude M of the angular momentum are given constants. These two equations, written in terms of the components of the vector $\mathbf{M}$, are

$$\frac{M_1^2}{I_1} + \frac{M_2^2}{I_2} + \frac{M_3^2}{I_3} = 2E, \tag{37.3}$$

$$M_1^2 + M_2^2 + M_3^2 = M^2. \tag{37.4}$$

From these equations we can already draw some conclusions concerning the nature of the motion. To do so, we notice that equations (37.3) and (37.4), regarded as involving co-ordinates M_1, M_2, M_3, are respectively the equation of an ellipsoid with semiaxes $\sqrt{(2EI_1)}$, $\sqrt{(2EI_2)}$, $\sqrt{(2EI_3)}$ and that of a sphere of radius M. When the vector $\mathbf{M}$ moves relative to the axes of inertia of the top, its terminus moves along the line of intersection of these two surfaces. Fig. 51 shows a number of such lines of intersection of an ellipsoid with spheres of various radii. The existence of an intersection is ensured by the obviously valid inequalities

$$2EI_1 < M^2 < 2EI_3, \tag{37.5}$$

which signify that the radius of the sphere (37.4) lies between the least and greatest semiaxes of the ellipsoid (37.3).

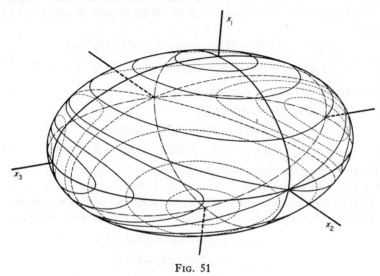

F_{IG}. 51

Let us examine the way in which these "paths"† of the terminus of the vector **M** change as M varies (for a given value of E). When M^2 is only slightly greater than $2EI_1$, the sphere intersects the ellipsoid in two small closed curves round the x_1-axis near the corresponding poles of the ellipsoid; as $M^2 \to 2EI_1$, these curves shrink to points at the poles. When M^2 increases, the curves become larger, and for $M^2 = 2EI_2$ they become two plane curves (ellipses) which intersect at the poles of the ellipsoid on the x_2-axis. When M^2 increases further, two separate closed paths again appear, but now round the poles on the x_3-axis; as $M^2 \to 2EI_3$ they shrink to points at these poles.

First of all, we may note that, since the paths are closed, the motion of the vector **M** relative to the top must be periodic; during one period the vector **M** describes some conical surface and returns to its original position.

Next, an essential difference in the nature of the paths near the various poles of the ellipsoid should be noted. Near the x_1 and x_3 axes, the paths lie entirely in the neighbourhood of the corresponding poles, but the paths which pass near the poles on the x_2-axis go elsewhere to great distances from those poles. This difference corresponds to a difference in the stability of the rotation of the top about its three axes of inertia. Rotation about the x_1 and x_3 axes (corresponding to the least and greatest of the three moments of inertia)

---

† The corresponding curves described by the terminus of the vector **Ω** are called *polhodes*.

is stable, in the sense that, if the top is made to deviate slightly from such a state, the resulting motion is close to the original one. A rotation about the x_2-axis, however, is unstable: a small deviation is sufficient to give rise to a motion which takes the top to positions far from its original one.

To determine the time dependence of the components of $\mathbf{\Omega}$ (or of the components of $\mathbf{M}$, which are proportional to those of $\mathbf{\Omega}$) we use Euler's equations (36.5). We express Ω_1 and Ω_3 in terms of Ω_2 by means of equations (37.2) and (37.3):

$$\begin{aligned}
\Omega_1{}^2 &= [(2EI_3 - M^2) - I_2(I_3 - I_2)\Omega_2{}^2]/I_1(I_3 - I_1), \\
\Omega_3{}^2 &= [(M^2 - 2EI_1) - I_2(I_2 - I_1)\Omega_2{}^2]/I_3(I_3 - I_1),
\end{aligned} \tag{37.6}$$

and substitute in the second equation (36.5), obtaining

$$\begin{aligned}
d\Omega_2/dt &= (I_3 - I_1)\Omega_1\Omega_3/I_2 \\
&= \sqrt{\{[(2EI_3 - M^2) - I_2(I_3 - I_2)\Omega_2{}^2] \times} \\
&\qquad \times [(M^2 - 2EI_1) - I_2(I_2 - I_1)\Omega_2{}^2]\}/I_2\sqrt{(I_1 I_3)}. \tag{37.7}
\end{aligned}$$

Integration of this equation gives the function $t(\Omega_2)$ as an elliptic integral. In reducing it to a standard form we shall suppose for definiteness that $M^2 > 2EI_2$; if this inequality is reversed, the suffixes 1 and 3 are interchanged in the following formulae. Using instead of t and Ω_2 the new variables

$$\begin{aligned}
\tau &= t\sqrt{[(I_3 - I_2)(M^2 - 2EI_1)/I_1 I_2 I_3]}, \\
s &= \Omega_2\sqrt{[I_2(I_3 - I_2)/(2EI_3 - M^2)]},
\end{aligned} \tag{37.8}$$

and defining a positive parameter $k^2 < 1$ by

$$k^2 = (I_2 - I_1)(2EI_3 - M^2)/(I_3 - I_2)(M^2 - 2EI_1), \tag{37.9}$$

we obtain

$$\tau = \int_0^s \frac{ds}{\sqrt{[(1 - s^2)(1 - k^2 s^2)]}},$$

the origin of time being taken at an instant when $\Omega_2 = 0$. When this integral is inverted we have a Jacobian elliptic function $s = \operatorname{sn}\tau$, and this gives Ω_2 as a function of time; $\Omega_1(t)$ and $\Omega_3(t)$ are algebraic functions of $\Omega_2(t)$ given by (37.6). Using the definitions $\operatorname{cn}\tau = \sqrt{(1 - \operatorname{sn}^2\tau)}$, $\operatorname{dn}\tau = \sqrt{(1 - k^2\operatorname{sn}^2\tau)}$, we find

$$\left.\begin{aligned}
\Omega_1 &= \sqrt{[(2EI_3 - M^2)/I_1(I_3 - I_1)]}\,\operatorname{cn}\tau, \\
\Omega_2 &= \sqrt{[(2EI_3 - M^2)/I_2(I_3 - I_2)]}\,\operatorname{sn}\tau, \\
\Omega_3 &= \sqrt{[(M^2 - 2EI_1)/I_3(I_3 - I_1)]}\,\operatorname{dn}\tau.
\end{aligned}\right\} \tag{37.10}$$

These are periodic functions, and their period in the variable τ is $4K$, where K is a complete elliptic integral of the first kind:

$$K = \int_0^1 \frac{ds}{\sqrt{[(1 - s^2)(1 - k^2 s^2)]}} = \int_0^{\frac{1}{2}\pi} \frac{du}{\sqrt{(1 - k^2\sin^2 u)}}. \tag{37.11}$$

The period in t is therefore

$$T = 4K\sqrt{[I_1 I_2 I_3/(I_3 - I_2)(M^2 - 2EI_1)]}. \tag{37.12}$$

After a time T the vector $\boldsymbol{\Omega}$ returns to its original position relative to the axes of the top. The top itself, however, does not return to its original position relative to the fixed system of co-ordinates; see below.

For $I_1 = I_2$, of course, formulae (37.10) reduce to those obtained in §36 for a symmetrical top: as $I_1 \to I_2$, the parameter $k^2 \to 0$, and the elliptic functions degenerate to circular functions: $\operatorname{sn} \tau \to \sin \tau$, $\operatorname{cn} \tau \to \cos \tau$, $\operatorname{dn} \tau \to 1$, and we return to formulae (36.7).

When $M^2 = 2EI_3$ we have $\Omega_1 = \Omega_2 = 0$, $\Omega_3 = $ constant, i.e. the vector $\boldsymbol{\Omega}$ is always parallel to the x_3-axis. This case corresponds to uniform rotation of the top about the x_3-axis. Similarly, for $M^2 = 2EI_1$ (when $\tau \equiv 0$) we have uniform rotation about the x_1-axis.

Let us now determine the absolute motion of the top in space (i.e. its motion relative to the fixed system of co-ordinates X, Y, Z). To do so, we use the Eulerian angles ψ, ϕ, θ, between the axes x_1, x_2, x_3 of the top and the axes X, Y, Z, taking the fixed Z-axis in the direction of the constant vector $\mathbf{M}$. Since the polar angle and azimuth of the Z-axis with respect to the axes x_1, x_2, x_3 are respectively θ and $\frac{1}{2}\pi - \psi$ (see the footnote to §35), we obtain on taking the components of $\mathbf{M}$ along the axes x_1, x_2, x_3

$$\left. \begin{aligned} M \sin\theta \sin\psi &= M_1 = I_1 \Omega_1, \\ M \sin\theta \cos\psi &= M_2 = I_2 \Omega_2, \\ M \cos\theta &= M_3 = I_3 \Omega_3. \end{aligned} \right\} \tag{37.13}$$

Hence

$$\cos\theta = I_3 \Omega_3/M, \qquad \tan\psi = I_1 \Omega_1/I_2 \Omega_2, \tag{37.14}$$

and from formulae (37.10)

$$\begin{aligned} \cos\theta &= \sqrt{[I_3(M^2 - 2EI_1)/M^2(I_3 - I_1)]} \operatorname{dn} \tau, \\ \tan\psi &= \sqrt{[I_1(I_3 - I_2)/I_2(I_3 - I_1)]} \operatorname{cn}\tau/\operatorname{sn}\tau, \end{aligned} \tag{37.15}$$

which give the angles θ and ψ as functions of time; like the components of the vector $\boldsymbol{\Omega}$, they are periodic functions, with period (37.12).

The angle ϕ does not appear in formulae (37.13), and to calculate it we must return to formulae (35.1), which express the components of $\boldsymbol{\Omega}$ in terms of the time derivatives of the Eulerian angles. Eliminating $\dot\theta$ from the equations $\Omega_1 = \dot\phi \sin\theta \sin\psi + \dot\theta \cos\psi$, $\Omega_2 = \dot\phi \sin\theta \cos\psi - \dot\theta \sin\psi$, we obtain $\dot\phi = (\Omega_1 \sin\psi + \Omega_2 \cos\psi)/\sin\theta$, and then, using formulae (37.13),

$$d\phi/dt = (I_1 \Omega_1^2 + I_2 \Omega_2^2)M/(I_1^2 \Omega_1^2 + I_2^2 \Omega_2^2). \tag{37.16}$$

The function $\phi(t)$ is obtained by integration, but the integrand involves elliptic functions in a complicated way. By means of some fairly complex transformations, the integral can be expressed in terms of theta functions; we shall not give the calculations,† but only the final result.

The function $\phi(t)$ can be represented (apart from an arbitrary additive constant) as a sum of two terms:

$$\phi(t) = \phi_1(t) + \phi_2(t), \tag{37.17}$$

one of which is given by

$$\exp[2i\phi_1(t)] = \vartheta_{01}\left(\frac{2t}{T} - i\alpha\right) \Big/ \vartheta_{01}\left(\frac{2t}{T} + i\alpha\right), \tag{37.18}$$

where ϑ_{01} is a theta function and α a real constant such that

$$\mathrm{sn}(2i\alpha K) = i\sqrt{[I_3(M^2 - 2EI_1)/I_1(2EI_3 - M^2)]}; \tag{37.19}$$

K and T are given by (37.11) and (37.12). The function on the right-hand side of (37.18) is periodic, with period $\frac{1}{2}T$, so that $\phi_1(t)$ varies by 2π during a time T. The second term in (37.17) is given by

$$\phi_2(t) = 2\pi t/T', \qquad \frac{1}{T'} = \frac{M}{2\pi I_1} - \frac{i}{\pi T}\frac{\vartheta_{01}'(i\alpha)}{\vartheta_{01}(i\alpha)}. \tag{37.20}$$

This function increases by 2π during a time T'. Thus the motion in ϕ is a combination of two periodic motions, one of the periods (T) being the same as the period of variation of the angles ψ and θ, while the other (T') is incommensurable with T. This incommensurability has the result that the top does not at any time return exactly to its original position.

PROBLEMS

PROBLEM 1. Determine the free rotation of a top about an axis near the x_3-axis or the x_1-axis.

SOLUTION. Let the x_3-axis be near the direction of **M**. Then the components M_1 and M_2 are small quantities, and the component $M_3 = M$ (apart from quantities of the second and higher orders of smallness). To the same accuracy the first two Euler's equations (36.5) can be written $dM_1/dt = \Omega_0 M_2(1 - I_3/I_2)$, $dM_2/dt = \Omega_0 M_1(I_3/I_1 - 1)$, where $\Omega_0 = M/I_3$. As usual we seek solutions for M_1 and M_2 proportional to $\exp(i\omega t)$, obtaining for the frequency ω

$$\omega = \Omega_0\sqrt{\left[\left(\frac{I_3}{I_1} - 1\right)\left(\frac{I_3}{I_2} - 1\right)\right]}. \tag{1}$$

The values of M_1 and M_2 are

$$M_1 = Ma\sqrt{\left(\frac{I_3}{I_2} - 1\right)}\cos\omega t, \qquad M_2 = Ma\sqrt{\left(\frac{I_3}{I_1} - 1\right)}\sin\omega t, \tag{2}$$

where a is an arbitrary small constant. These formulae give the motion of the vector **M** relative to the top. In Fig. 51, the terminus of the vector **M** describes, with frequency ω, a small ellipse about the pole on the x_3-axis.

To determine the absolute motion of the top in space, we calculate its Eulerian angles. In the present case the angle θ between the x_3-axis and the Z-axis (direction of **M**) is small,

---

† These are given by E. T. WHITTAKER, *A Treatise on the Analytical Dynamics of Particles and Rigid Bodies*, 4th ed., Chapter VI, Dover, New York 1944.

and by formulae (37.14) $\tan \psi = M_1/M_2$, $\theta^2 \approx 2(1-\cos\theta) = 2(1-M_3/M) \approx (M_1^2+M_2^2)/M^2$; substituting (2), we obtain

$$\tan \psi = \sqrt{[I_1(I_3-I_2)/I_2(I_3-I_1)]}\, \cot \omega t,$$

$$\theta^2 = a^2\left[\left(\frac{I_3}{I_2}-1\right)\cos^2\omega t + \left(\frac{I_3}{I_1}-1\right)\sin^2\omega t\right]. \tag{3}$$

To find ϕ, we note that, by the third formula (35.1), we have, for $\theta \ll 1$, $\Omega_0 \approx \Omega_3 \approx \dot\psi + \dot\phi$. Hence

$$\phi = \Omega_0 t - \psi, \tag{4}$$

omitting an arbitrary constant of integration.

A clearer idea of the nature of the motion of the top is obtained if we consider the change in direction of the three axes of inertia. Let $\mathbf{n}_1, \mathbf{n}_2, \mathbf{n}_3$ be unit vectors along these axes. The vectors $\mathbf{n}_1$ and $\mathbf{n}_2$ rotate uniformly in the XY-plane with frequency Ω_0, and at the same time execute small transverse oscillations with frequency ω. These oscillations are given by the Z-components of the vectors:

$$n_{1z} \approx M_1/M = a\sqrt{(I_3/I_2-1)}\,\cos \omega t,$$
$$n_{2z} \approx M_2/M = a\sqrt{(I_3/I_1-1)}\,\sin \omega t.$$

For the vector $\mathbf{n}_3$ we have, to the same accuracy, $n_{3x} \approx \theta \sin\phi$, $n_{3y} \approx -\theta \cos\phi$, $n_{3z} \approx 1$. (The polar angle and azimuth of $\mathbf{n}_3$ with respect to the axes X, Y, Z are θ and $\phi - \frac{1}{2}\pi$; see the footnote to §35.) We also write, using formulae (37.13),

$$\begin{aligned}
n_{3x} &= \theta \sin(\Omega_0 t - \psi)\\
&= \theta \sin \Omega_0 t \cos \psi - \theta \cos \Omega_0 t \sin \psi\\
&= (M_2/M) \sin \Omega_0 t - (M_1/M) \cos \Omega_0 t\\
&= a\sqrt{\left(\frac{I_3}{I_1}-1\right)} \sin \Omega_0 t \sin \omega t - a\sqrt{\left(\frac{I_3}{I_2}-1\right)} \cos \Omega_0 t \cos \omega t\\
&= -\tfrac{1}{2}a\left[\sqrt{\left(\frac{I_3}{I_1}-1\right)} + \sqrt{\left(\frac{I_3}{I_2}-1\right)}\right]\cos(\Omega_0+\omega)t +\\
&\quad +\tfrac{1}{2}a\left[\sqrt{\left(\frac{I_3}{I_1}-1\right)} - \sqrt{\left(\frac{I_3}{I_2}-1\right)}\right]\cos(\Omega_0-\omega)t.
\end{aligned}$$

Similarly

$$\begin{aligned}
n_{3y} &= -\tfrac{1}{2}a\left[\sqrt{\left(\frac{I_3}{I_1}-1\right)} + \sqrt{\left(\frac{I_3}{I_2}-1\right)}\right]\sin(\Omega_0+\omega)t +\\
&\quad +\tfrac{1}{2}a\left[\sqrt{\left(\frac{I_3}{I_1}-1\right)} - \sqrt{\left(\frac{I_3}{I_2}-1\right)}\right]\sin(\Omega_0-\omega)t.
\end{aligned}$$

From this we see that the motion of $\mathbf{n}_3$ is a superposition of two rotations about the Z-axis with frequencies $\Omega_0 \pm \omega$.

PROBLEM 2. Determine the free rotation of a top for which $M^2 = 2EI_2$.

SOLUTION. This case corresponds to the movement of the terminus of $\mathbf{M}$ along a curve through the pole on the x_2-axis (Fig. 51). Equation (37.7) becomes $ds/d\tau = 1-s^2$, $\tau = t\sqrt{[(I_2-I_1)(I_3-I_2)/I_1I_3]}\Omega_0$, $s = \Omega_2/\Omega_0$, where $\Omega_0 = M/I_2 = 2E/M$. Integration of this equation and the use of formulae (37.6) gives

$$\left.\begin{aligned}
\Omega_1 &= \Omega_0\sqrt{[I_2(I_3-I_2)/I_1(I_3-I_1)]}\,\text{sech}\,\tau,\\
\Omega_2 &= \Omega_0 \tanh \tau,\\
\Omega_3 &= \Omega_0\sqrt{[I_2(I_2-I_1)/I_3(I_3-I_1)]}\,\text{sech}\,\tau.
\end{aligned}\right\} \tag{1}$$

To describe the absolute motion of the top, we use Eulerian angles, defining θ as the angle between the Z-axis (direction of $\mathbf{M}$) and the x_2-axis (not the x_3-axis as previously). In formulae (37.14) and (37.16), which relate the components of the vector $\mathbf{\Omega}$ to the Eulerian angles, we must cyclically permute the suffixes 1, 2, 3 to 3, 1, 2. Substitution of (1) in these formulae then gives $\cos \theta = \tanh \tau$, $\phi = \Omega_0 t + \text{constant}$, $\tan \psi = \sqrt{[I_3(I_2 - I_1)/I_1(I_3 - I_2)]}$.

It is seen from these formulae that, as $t \to \infty$, the vector $\boldsymbol{\Omega}$ asymptotically approaches the x_2-axis, which itself asymptotically approaches the Z-axis.

§38. Rigid bodies in contact

The equations of motion (34.1) and (34.3) show that the conditions of equilibrium for a rigid body can be written as the vanishing of the total force and total torque on the body:

$$\mathbf{F} = \sum \mathbf{f} = 0, \qquad \mathbf{K} = \sum \mathbf{r} \times \mathbf{f} = 0. \tag{38.1}$$

Here the summation is over all the external forces acting on the body, and $\mathbf{r}$ is the radius vector of the "point of application"; the origin with respect to which the torque is defined may be chosen arbitrarily, since if $\mathbf{F} = 0$ the value of $\mathbf{K}$ does not depend on this choice (see (34.5)).

If we have a system of rigid bodies in contact, the conditions (38.1) for each body separately must hold in equilibrium. The forces considered must include those exerted on each body by those with which it is in contact. These forces at the points of contact are called *reactions*. It is obvious that the mutual reactions of any two bodies are equal in magnitude and opposite in direction.

In general, both the magnitudes and the directions of the reactions are found by solving simultaneously the equations of equilibrium (38.1) for all the bodies. In some cases, however, their directions are given by the conditions of the problem. For example, if two bodies can slide freely on each other, the reaction between them is normal to the surface.

If two bodies in contact are in relative motion, dissipative forces of *friction* arise, in addition to the reaction.

There are two possible types of motion of bodies in contact—*sliding* and *rolling*. In sliding, the reaction is perpendicular to the surfaces in contact, and the friction is tangential. Pure rolling, on the other hand, is characterised by the fact that there is no relative motion of the bodies at the point of contact; that is, a rolling body is at every instant as it were fixed to the point of contact. The reaction may be in any direction, i.e. it need not be normal to the surfaces in contact. The friction in rolling appears as an additional torque which opposes rolling.

If the friction in sliding is negligibly small, the surfaces concerned are said to be *perfectly smooth*. If, on the other hand, only pure rolling without sliding is possible, and the friction in rolling can be neglected, the surfaces are said to be *perfectly rough*.

In both these cases the frictional forces do not appear explicitly in the problem, which is therefore purely one of mechanics. If, on the other hand, the properties of the friction play an essential part in determining the motion, then the latter is not a purely mechanical process (cf. §25).

Contact between two bodies reduces the number of their degrees of freedom as compared with the case of free motion. Hitherto, in discussing such problems, we have taken this reduction into account by using co-ordinates which correspond directly to the actual number of degrees of freedom. In rolling, however, such a choice of co-ordinates may be impossible.

The condition imposed on the motion of rolling bodies is that the velocities of the points in contact should be equal; for example, when a body rolls on a fixed surface, the velocity of the point of contact must be zero. In the general case, this condition is expressed by the *equations of constraint*, of the form

$$\sum_i c_{\alpha i} \dot{q}_i = 0, \tag{38.2}$$

where the $c_{\alpha i}$ are functions of the co-ordinates only, and the suffix α denumerates the equations. If the left-hand sides of these equations are not the total time derivatives of some functions of the co-ordinates, the equations cannot be integrated. In other words, they cannot be reduced to relations between the co-ordinates only, which could be used to express the position of the bodies in terms of fewer co-ordinates, corresponding to the actual number of degrees of freedom. Such constraints are said to be *non-holonomic*, as opposed to *holonomic* constraints, which impose relations between the co-ordinates only.

Let us consider, for example, the rolling of a sphere on a plane. As usual, we denote by $\mathbf{V}$ the translational velocity (the velocity of the centre of the sphere), and by $\boldsymbol{\Omega}$ the angular velocity of rotation. The velocity of the point of contact with the plane is found by putting $\mathbf{r} = -a\mathbf{n}$ in the general formula $\mathbf{v} = \mathbf{V} + \boldsymbol{\Omega} \times \mathbf{r}$; a is the radius of the sphere and $\mathbf{n}$ a unit vector along the normal to the plane. The required condition is that there should be no sliding at the point of contact, i.e.

$$\mathbf{V} - a\boldsymbol{\Omega} \times \mathbf{n} = 0. \tag{38.3}$$

This cannot be integrated: although the velocity $\mathbf{V}$ is the total time derivative of the radius vector of the centre of the sphere, the angular velocity is not in general the total time derivative of any co-ordinate. The constraint (38.3) is therefore non-holonomic.[†]

Since the equations of non-holonomic constraints cannot be used to reduce the number of co-ordinates, when such constraints are present it is necessary to use co-ordinates which are not all independent. To derive the corresponding Lagrange's equations, we return to the principle of least action.

The existence of the constraints (38.2) places certain restrictions on the possible values of the variations of the co-ordinates: multiplying equations (38.2) by δt, we find that the variations δq_i are not independent, but are related by

$$\sum_i c_{\alpha i} \delta q_i = 0. \tag{38.4}$$

---

† It may be noted that the similar constraint in the rolling of a cylinder *is* holonomic. In that case the axis of rotation has a fixed direction in space, and hence $\Omega = \mathrm{d}\phi/\mathrm{d}t$ is the total derivative of the angle ϕ of rotation of the cylinder about its axis. The condition (38.3) can therefore be integrated, and gives a relation between the angle ϕ and the co-ordinate of the centre of mass.

This must be taken into account in varying the action. According to Lagrange's method of finding conditional extrema, we must add to the integrand in the variation of the action

$$\delta S = \int \sum_i \delta q_i \left[\frac{\partial L}{\partial q_i} - \frac{\mathrm{d}}{\mathrm{d}t}\left(\frac{\partial L}{\partial \dot{q}_i} \right) \right] \mathrm{d}t$$

the left-hand sides of equations (38.4) multiplied by undetermined coefficients λ_α (functions of the co-ordinates), and then equate the integral to zero. In so doing the variations δq_i are regarded as entirely independent, and the result is

$$\frac{\mathrm{d}}{\mathrm{d}t}\left(\frac{\partial L}{\partial \dot{q}_i} \right) - \frac{\partial L}{\partial q_i} = \sum_\alpha \lambda_\alpha c_{\alpha i}. \tag{38.5}$$

These equations, together with the constraint equations (38.2), form a complete set of equations for the unknowns q_i and λ_α.

The reaction forces do not appear in this treatment, and the contact of the bodies is fully allowed for by means of the constraint equations. There is, however, another method of deriving the equations of motion for bodies in contact, in which the reactions are introduced explicitly. The essential feature of this method, which is sometimes called *d'Alembert's principle*, is to write for each of the bodies in contact the equations.

$$\mathrm{d}\mathbf{P}/\mathrm{d}t = \sum \mathbf{f}, \qquad \mathrm{d}\mathbf{M}/\mathrm{d}t = \sum \mathbf{r} \times \mathbf{f}, \tag{38.6}$$

wherein the forces $\mathbf{f}$ acting on each body include the reactions. The latter are initially unknown and are determined, together with the motion of the body, by solving the equations. This method is equally applicable for both holonomic and non-holonomic constraints.

PROBLEMS

PROBLEM 1. Using d'Alembert's principle, find the equations of motion of a homogeneous sphere rolling on a plane under an external force $\mathbf{F}$ and torque $\mathbf{K}$.

SOLUTION. The constraint equation is (38.3). Denoting the reaction force at the point of contact between the sphere and the plane by $\mathbf{R}$, we have equations (38.6) in the form

$$\mu \, \mathrm{d}\mathbf{V}/\mathrm{d}t = \mathbf{F} + \mathbf{R}, \tag{1}$$

$$I \, \mathrm{d}\boldsymbol{\Omega}/\mathrm{d}t = \mathbf{K} - a\mathbf{n} \times \mathbf{R}, \tag{2}$$

where we have used the facts that $\mathbf{P} = \mu\mathbf{V}$ and, for a spherical top, $\mathbf{M} = I\boldsymbol{\Omega}$. Differentiating the constraint equation (38.3) with respect to time, we have $\dot{\mathbf{V}} = a\dot{\boldsymbol{\Omega}} \times \mathbf{n}$. Substituting in equation (1) and eliminating $\dot{\boldsymbol{\Omega}}$ by means of (2), we obtain $(I/a\mu)(\mathbf{F}+\mathbf{R}) = \mathbf{K}\times\mathbf{n} - a\mathbf{R} + a\mathbf{n}(\mathbf{n}\cdot\mathbf{R})$, which relates $\mathbf{R}$, $\mathbf{F}$ and $\mathbf{K}$. Writing this equation in components and substituting $I = \frac{2}{5}\mu a^2$ (§32, Problem 2(b)), we have

$$R_x = \frac{5}{7a}K_y - \frac{2}{7}F_x, \qquad R_y = -\frac{5}{7a}K_x - \frac{2}{7}F_y, \qquad R_z = -F_z,$$

where the plane is taken as the xy-plane. Finally, substituting these expressions in (1), we obtain the equations of motion involving only the given external force and torque:

$$\frac{dV_x}{dt} = \frac{5}{7\mu}\left(F_x + \frac{K_y}{a}\right),$$

$$\frac{dV_y}{dt} = \frac{5}{7\mu}\left(F_y - \frac{K_x}{a}\right).$$

The components Ω_x, Ω_y of the angular velocity are given in terms of V_x, V_y by the constraint equation (38.3); for Ω_z we have the equation $\frac{2}{5}\mu a^2 \, d\Omega_z/dt = K_z$, the z-component of equation (2).

PROBLEM 2. A uniform rod BD of weight P and length l rests against a wall as shown in Fig. 52 and its lower end B is held by a string AB. Find the reaction of the wall and the tension in the string.

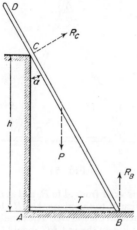

FIG. 52

SOLUTION. The weight of the rod can be represented by a force P vertically downwards, applied at its midpoint. The reactions R_B and R_C are respectively vertically upwards and perpendicular to the rod; the tension T in the string is directed from B to A. The solution of the equations of equilibrium gives $R_C = (Pl/4h)\sin 2\alpha$, $R_B = P - R_C \sin\alpha$, $T = R_C \cos\alpha$.

PROBLEM 3. A rod of weight P has one end A on a vertical plane and the other end B on a horizontal plane (Fig. 53), and is held in position by two horizontal strings AD and BC,

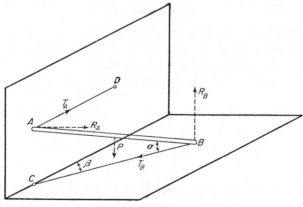

FIG. 53

the latter being in the same vertical plane as AB. Determine the reactions of the planes and the tensions in the strings.

SOLUTION. The tensions T_A and T_B are from A to D and from B to C respectively. The reactions R_A and R_B are perpendicular to the corresponding planes. The solution of the equations of equilibrium gives $R_B = P$, $T_B = \frac{1}{2}P \cot \alpha$, $R_A = T_B \sin \beta$, $T_A = T_B \cos \beta$.

PROBLEM 4. Two rods of length l and negligible weight are hinged together, and their ends are connected by a string AB (Fig. 54). They stand on a plane, and a force F is applied at the midpoint of one rod. Determine the reactions.

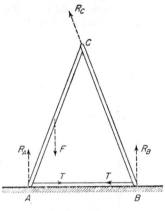

FIG. 54

SOLUTION. The tension T acts at A from A to B, and at B from B to A. The reactions R_A and R_B at A and B are perpendicular to the plane. Let R_C be the reaction on the rod AC at the hinge; then a reaction $-R_C$ acts on the rod BC. The condition that the sum of the moments of the forces R_B, T and $-R_C$ acting on the rod BC should be zero shows that R_C acts along BC. The remaining conditions of equilibrium (for the two rods separately) give $R_A = \frac{3}{4}F$, $R_B = \frac{1}{4}F$, $R_C = \frac{1}{4}F$ cosec α, $T = \frac{1}{4}F \cot \alpha$, where α is the angle CAB.

§39. Motion in a non-inertial frame of reference

Up to this point we have always used inertial frames of reference in discussing the motion of mechanical systems. For example, the Lagrangian

$$L_0 = \tfrac{1}{2}mv_0^2 - U, \tag{39.1}$$

and the corresponding equation of motion $m \, dv_0/dt = -\partial U/\partial r$, for a single particle in an external field are valid only in an inertial frame. (In this section the suffix 0 denotes quantities pertaining to an inertial frame.)

Let us now consider what the equations of motion will be in a non-inertial frame of reference. The basis of the solution of this problem is again the principle of least action, whose validity does not depend on the frame of reference chosen. Lagrange's equations

$$\frac{d}{dt}\left(\frac{\partial L}{\partial v}\right) = \frac{\partial L}{\partial r} \tag{39.2}$$

are likewise valid, but the Lagrangian is no longer of the form (39.1), and to derive it we must carry out the necessary transformation of the function L_0.

This transformation is done in two steps. Let us first consider a frame of reference K' which moves with a translational velocity $\mathbf{V}(t)$ relative to the inertial frame K_0. The velocities $\mathbf{v}_0$ and $\mathbf{v}'$ of a particle in the frames K_0 and K' respectively are related by

$$\mathbf{v}_0 = \mathbf{v}' + \mathbf{V}(t). \tag{39.3}$$

Substitution of this in (39.1) gives the Lagrangian in K':

$$L' = \tfrac{1}{2}m\mathbf{v}'^2 + m\mathbf{v}' \cdot \mathbf{V} + \tfrac{1}{2}m\mathbf{V}^2 - U.$$

Now $\mathbf{V}^2(t)$ is a given function of time, and can be written as the total derivative with respect to t of some other function; the third term in L' can therefore be omitted. Next, $\mathbf{v}' = d\mathbf{r}'/dt$, where $\mathbf{r}'$ is the radius vector of the particle in the frame K'. Hence

$$m\mathbf{V}(t)\cdot\mathbf{v}' = m\mathbf{V}\cdot d\mathbf{r}'/dt = d(m\mathbf{V}\cdot\mathbf{r}')/dt - m\mathbf{r}'\cdot d\mathbf{V}/dt.$$

Substituting in the Lagrangian and again omitting the total time derivative, we have finally

$$L' = \tfrac{1}{2}mv'^2 - m\mathbf{W}(t)\cdot\mathbf{r}' - U, \tag{39.4}$$

where $\mathbf{W} = d\mathbf{V}/dt$ is the translational acceleration of the frame K'.
    The Lagrange's equation derived from (39.4) is

$$m\frac{d\mathbf{v}'}{dt} = -\frac{\partial U}{\partial \mathbf{r}'} - m\mathbf{W}(t). \tag{39.5}$$

Thus an accelerated translational motion of a frame of reference is equivalent, as regards its effect on the equations of motion of a particle, to the application of a uniform field of force equal to the mass of the particle multiplied by the acceleration $\mathbf{W}$, in the direction opposite to this acceleration.
    Let us now bring in a further frame of reference K, whose origin coincides with that of K', but which rotates relative to K' with angular velocity $\boldsymbol{\Omega}(t)$. Thus K executes both a translational and a rotational motion relative to the inertial frame K_0.
    The velocity $\mathbf{v}'$ of the particle relative to K' is composed of its velocity $\mathbf{v}$ relative to K and the velocity $\boldsymbol{\Omega}\times\mathbf{r}$ of its rotation with K: $\mathbf{v}' = \mathbf{v} + \boldsymbol{\Omega}\times\mathbf{r}$ (since the radius vectors $\mathbf{r}$ and $\mathbf{r}'$ in the frames K and K' coincide). Substituting this in the Lagrangian (39.4), we obtain

$$L = \tfrac{1}{2}mv^2 + m\mathbf{v}\cdot\boldsymbol{\Omega}\times\mathbf{r} + \tfrac{1}{2}m(\boldsymbol{\Omega}\times\mathbf{r})^2 - m\mathbf{W}\cdot\mathbf{r} - U. \tag{39.6}$$

This is the general form of the Lagrangian of a particle in an arbitrary, not necessarily inertial, frame of reference. The rotation of the frame leads to the appearance in the Lagrangian of a term linear in the velocity of the particle.
    To calculate the derivatives appearing in Lagrange's equation, we write

the total differential

$$dL = m\mathbf{v}\cdot d\mathbf{v} + md\mathbf{v}\cdot\mathbf{\Omega}\times\mathbf{r} + m\mathbf{v}\cdot\mathbf{\Omega}\times d\mathbf{r} +$$
$$+ m(\mathbf{\Omega}\times\mathbf{r})\cdot(\mathbf{\Omega}\times d\mathbf{r}) - m\mathbf{W}\cdot d\mathbf{r} - (\partial U/\partial\mathbf{r})\cdot d\mathbf{r}$$
$$= m\mathbf{v}\cdot d\mathbf{v} + md\mathbf{v}\cdot\mathbf{\Omega}\times\mathbf{r} + md\mathbf{r}\cdot\mathbf{v}\times\mathbf{\Omega} +$$
$$+ m(\mathbf{\Omega}\times\mathbf{r})\times\mathbf{\Omega}\cdot d\mathbf{r} - m\mathbf{W}\cdot d\mathbf{r} - (\partial U/\partial\mathbf{r})\cdot d\mathbf{r}.$$

The terms in $d\mathbf{v}$ and $d\mathbf{r}$ give

$$\partial L/\partial\mathbf{v} = m\mathbf{v} + m\mathbf{\Omega}\times\mathbf{r},$$
$$\partial L/\partial\mathbf{r} = m\mathbf{v}\times\mathbf{\Omega} + m(\mathbf{\Omega}\times\mathbf{r})\times\mathbf{\Omega} - m\mathbf{W} - \partial U/\partial\mathbf{r}.$$

Substitution of these expressions in (39.2) gives the required equation of motion:

$$md\mathbf{v}/dt = -\partial U/\partial\mathbf{r} - m\mathbf{W} + m\mathbf{r}\times\dot{\mathbf{\Omega}} + 2m\mathbf{v}\times\mathbf{\Omega} + m\mathbf{\Omega}\times(\mathbf{r}\times\mathbf{\Omega}). \quad (39.7)$$

We see that the "inertia forces" due to the rotation of the frame consist of three terms. The force $m\mathbf{r}\times\dot{\mathbf{\Omega}}$ is due to the non-uniformity of the rotation, but the other two terms appear even if the rotation is uniform. The force $2m\mathbf{v}\times\mathbf{\Omega}$ is called the *Coriolis force*; unlike any other (non-dissipative) force hitherto considered, it depends on the velocity of the particle. The force $m\mathbf{\Omega}\times(\mathbf{r}\times\mathbf{\Omega})$ is called the *centrifugal force*. It lies in the plane through $\mathbf{r}$ and $\mathbf{\Omega}$, is perpendicular to the axis of rotation (i.e. to $\mathbf{\Omega}$), and is directed away from the axis. The magnitude of this force is $m\rho\Omega^2$, where ρ is the distance of the particle from the axis of rotation.

Let us now consider the particular case of a uniformly rotating frame with no translational acceleration. Putting in (39.6) and (39.7) $\mathbf{\Omega} = $ constant, $\mathbf{W} = 0$, we obtain the Lagrangian

$$L = \tfrac{1}{2}mv^2 + m\mathbf{v}\cdot\mathbf{\Omega}\times\mathbf{r} + \tfrac{1}{2}m(\mathbf{\Omega}\times\mathbf{r})^2 - U \qquad (39.8)$$

and the equation of motion

$$md\mathbf{v}/dt = -\partial U/\partial\mathbf{r} + 2m\mathbf{v}\times\mathbf{\Omega} + m\mathbf{\Omega}\times(\mathbf{r}\times\mathbf{\Omega}). \qquad (39.9)$$

The energy of the particle in this case is obtained by substituting

$$\mathbf{p} = \partial L/\partial\mathbf{v} = m\mathbf{v} + m\mathbf{\Omega}\times\mathbf{r} \qquad (39.10)$$

in $E = \mathbf{p}\cdot\mathbf{v} - L$, which gives

$$E = \tfrac{1}{2}mv^2 - \tfrac{1}{2}m(\mathbf{\Omega}\times\mathbf{r})^2 + U. \qquad (39.11)$$

It should be noticed that the energy contains no term linear in the velocity. The rotation of the frame simply adds to the energy a term depending only on the co-ordinates of the particle and proportional to the square of the angular velocity. This additional term $-\tfrac{1}{2}m(\mathbf{\Omega}\times\mathbf{r})^2$ is called the *centrifugal potential energy*.

The velocity $\mathbf{v}$ of the particle relative to the uniformly rotating frame of reference is related to its velocity $\mathbf{v}_0$ relative to the inertial frame K_0 by

$$\mathbf{v}_0 = \mathbf{v} + \mathbf{\Omega}\times\mathbf{r}. \qquad (39.12)$$

The momentum $\mathbf{p}$ (39.10) of the particle in the frame K is therefore the same as its momentum $\mathbf{p_0} = m\mathbf{v_0}$ in the frame K_0. The angular momenta $\mathbf{M_0} = \mathbf{r} \times \mathbf{p_0}$ and $\mathbf{M} = \mathbf{r} \times \mathbf{p}$ are likewise equal. The energies of the particle in the two frames are not the same, however. Substituting $\mathbf{v}$ from (39.12) in (39.11), we obtain $E = \frac{1}{2}mv_0^2 - m\mathbf{v_0} \cdot \mathbf{\Omega} \times \mathbf{r} + U = \frac{1}{2}mv_0^2 + U - m\mathbf{r} \times \mathbf{v_0} \cdot \mathbf{\Omega}$. The first two terms are the energy E_0 in the frame K_0. Using the angular momentum $\mathbf{M}$, we have

$$E = E_0 - \mathbf{M} \cdot \mathbf{\Omega}. \tag{39.13}$$

This formula gives the law of transformation of energy when we change to a uniformly rotating frame. Although it has been derived for a single particle, the derivation can evidently be generalised immediately to any system of particles, and the same formula (39.13) is obtained.

PROBLEMS

PROBLEM 1. Find the deflection of a freely falling body from the vertical caused by the Earth's rotation, assuming the angular velocity of this rotation to be small.

SOLUTION. In a gravitational field $U = -m\mathbf{g} \cdot \mathbf{r}$, where $\mathbf{g}$ is the gravity acceleration vector; neglecting the centrifugal force in equation (39.9) as containing the square of $\mathbf{\Omega}$, we have the equation of motion

$$\dot{\mathbf{v}} = 2\mathbf{v} \times \mathbf{\Omega} + \mathbf{g}. \tag{1}$$

This equation may be solved by successive approximations. To do so, we put $\mathbf{v} = \mathbf{v_1} + \mathbf{v_2}$, where $\mathbf{v_1}$ is the solution of the equation $\dot{\mathbf{v}}_1 = \mathbf{g}$, i.e. $\mathbf{v_1} = \mathbf{g}t + \mathbf{v_0}$ ($\mathbf{v_0}$ being the initial velocity). Substituting $\mathbf{v} = \mathbf{v_1} + \mathbf{v_2}$ in (1) and retaining only $\mathbf{v_1}$ on the right, we have for $\mathbf{v_2}$ the equation $\dot{\mathbf{v}}_2 = 2\mathbf{v_1} \times \mathbf{\Omega} = 2t\mathbf{g} \times \mathbf{\Omega} + 2\mathbf{v_0} \times \mathbf{\Omega}$. Integration gives

$$\mathbf{r} = \mathbf{h} + \mathbf{v_0}t + \frac{1}{2}\mathbf{g}t^2 + \frac{1}{3}t^3\mathbf{g} \times \mathbf{\Omega} + t^2\mathbf{v_0} \times \mathbf{\Omega}, \tag{2}$$

where $\mathbf{h}$ is the initial radius vector of the particle.

Let the z-axis be vertically upwards, and the x-axis towards the pole; then $g_x = g_y = 0$, $g_z = -g$; $\Omega_x = \Omega \cos\lambda$, $\Omega_y = 0$, $\Omega_z = \Omega \sin\lambda$, where λ is the latitude (which for definiteness we take to be north). Putting $\mathbf{v_0} = 0$ in (2), we find $x = 0$, $y = -\frac{1}{3}t^3g\Omega\cos\lambda$. Substitution of the time of fall $t \approx \sqrt{(2h/g)}$ gives finally $x = 0$, $y = -\frac{1}{3}(2h/g)^{3/2}g\Omega\cos\lambda$, the negative value indicating an eastward deflection.

PROBLEM 2. Determine the deflection from coplanarity of the path of a particle thrown from the Earth's surface with velocity $\mathbf{v_0}$.

SOLUTION. Let the xz-plane be such as to contain the velocity $\mathbf{v_0}$. The initial altitude $h = 0$. The lateral deviation is given by (2), Problem 1: $y = -\frac{1}{3}t^3g\Omega_x + t^2(\Omega_x v_{0z} - \Omega_z v_{0x})$ or, substituting the time of flight $t \approx 2v_{0z}/g$, $y = 4v_{0z}^2(\frac{1}{3}v_{0z}\Omega_x - v_{0x}\Omega_z)/g^2$.

PROBLEM 3. Determine the effect of the Earth's rotation on small oscillations of a pendulum (the problem of *Foucault's pendulum*).

SOLUTION. Neglecting the vertical displacement of the pendulum, as being a quantity of the second order of smallness, we can regard the motion as taking place in the horizontal xy-plane. Omitting terms in Ω^2, we have the equations of motion $\ddot{x} + \omega^2 x = 2\Omega_z\dot{y}$, $\ddot{y} + \omega^2 y = -2\Omega_z\dot{x}$, where ω is the frequency of oscillation of the pendulum if the Earth's rotation is neglected. Multiplying the second equation by i and adding, we obtain a single equation

$\ddot{\xi}+2i\Omega_z\dot{\xi}+\omega^2\xi = 0$ for the complex quantity $\xi = x+iy$. For $\Omega_z \ll \omega$, the solution of this equation is

$$\xi = \exp(-i\Omega_z t)[A_1 \exp(i\omega t)+A_2 \exp(-i\omega t)]$$

or

$$x+iy = (x_0+iy_0)\, \exp(-i\Omega_z t),$$

where the functions $x_0(t)$, $y_0(t)$ give the path of the pendulum when the Earth's rotation is neglected. The effect of this rotation is therefore to turn the path about the vertical with angular velocity Ω_z.

CHAPTER VII

THE CANONICAL EQUATIONS†

§40. Hamilton's equations

THE formulation of the laws of mechanics in terms of the Lagrangian, and of Lagrange's equations derived from it, presupposes that the mechanical state of a system is described by specifying its generalised co-ordinates and velocities. This is not the only possible mode of description, however. A number of advantages, especially in the study of certain general problems of mechanics, attach to a description in terms of the generalised co-ordinates and momenta of the system. The question therefore arises of the form of the equations of motion corresponding to that formulation of mechanics.

The passage from one set of independent variables to another can be effected by means of what is called in mathematics *Legendre's transformation*. In the present case this transformation is as follows. The total differential of the Lagrangian as a function of co-ordinates and velocities is

$$dL = \sum_i \frac{\partial L}{\partial q_i}\,dq_i + \sum_i \frac{\partial L}{\partial \dot{q}_i}\,d\dot{q}_i.$$

This expression may be written

$$dL = \sum \dot{p}_i\,dq_i + \sum p_i\,d\dot{q}_i, \tag{40.1}$$

since the derivatives $\partial L/\partial \dot{q}_i$ are, by definition, the generalised momenta, and $\partial L/\partial q_i = \dot{p}_i$ by Lagrange's equations. Writing the second term in (40.1) as $\sum p_i\,d\dot{q}_i = d(\sum p_i \dot{q}_i) - \sum \dot{q}_i\,dp_i$, taking the differential $d(\sum p_i \dot{q}_i)$ to the left-hand side, and reversing the signs, we obtain from (40.1)

$$d\left(\sum p_i \dot{q}_i - L\right) = -\sum \dot{p}_i\,dq_i + \sum \dot{q}_i\,dp_i.$$

The argument of the differential is the energy of the system (cf. §6); expressed in terms of co-ordinates and momenta, it is called the *Hamilton's function* or *Hamiltonian* of the system:

$$H(p, q, t) = \sum_i p_i \dot{q}_i - L. \tag{40.2}$$

---

† The reader may find useful the following table showing certain differences between the nomenclature used in this book and that which is generally used in the English literature.

Here	*Elsewhere*
Principle of least action	Hamilton's principle
Maupertuis' principle	{ Principle of least action { Maupertuis' principle
Action	Hamilton's principal function
Abbreviated action	Action
—*Translators.*	

From the equation in differentials

$$dH = -\sum \dot{p}_i \, dq_i + \sum \dot{q}_i \, dp_i, \tag{40.3}$$

in which the independent variables are the co-ordinates and momenta, we have the equations

$$\dot{q}_i = \partial H / \partial p_i, \qquad \dot{p}_i = -\partial H / \partial q_i. \tag{40.4}$$

These are the required equations of motion in the variables p and q, and are called *Hamilton's equations*. They form a set of $2s$ first-order differential equations for the $2s$ unknown functions $p_i(t)$ and $q_i(t)$, replacing the s second-order equations in the Lagrangian treatment. Because of their simplicity and symmetry of form, they are also called *canonical equations*.

The total time derivative of the Hamiltonian is

$$\frac{dH}{dt} = \frac{\partial H}{\partial t} + \sum \frac{\partial H}{\partial q_i} \dot{q}_i + \sum \frac{\partial H}{\partial p_i} \dot{p}_i.$$

Substitution of $\dot{q}_i$ and $\dot{p}_i$ from equations (40.4) shows that the last two terms cancel, and so

$$dH/dt = \partial H/\partial t. \tag{40.5}$$

In particular, if the Hamiltonian does not depend explicitly on time, then $dH/dt = 0$, and we have the law of conservation of energy.

As well as the dynamical variables q, $\dot{q}$ or q, p, the Lagrangian and the Hamiltonian involve various parameters which relate to the properties of the mechanical system itself, or to the external forces on it. Let λ be one such parameter. Regarding it as a variable, we have instead of (40.1)

$$dL = \sum \dot{p}_i \, dq_i + \sum p_i \, d\dot{q}_i + (\partial L/\partial \lambda) \, d\lambda,$$

and (40.3) becomes

$$dH = -\sum \dot{p}_i \, dq_i + \sum \dot{q}_i \, dp_i - (\partial L/\partial \lambda) \, d\lambda.$$

Hence

$$(\partial H/\partial \lambda)_{p,q} = -(\partial L/\partial \lambda)_{\dot{q},q}, \tag{40.6}$$

which relates the derivatives of the Lagrangian and the Hamiltonian with respect to the parameter λ. The suffixes to the derivatives show the quantities which are to be kept constant in the differentiation.

This result can be put in another way. Let the Lagrangian be of the form $L = L_0 + L'$, where L' is a small correction to the function L_0. Then the corresponding addition H' in the Hamiltonian $H = H_0 + H'$ is related to L' by

$$(H')_{p,q} = -(L')_{\dot{q} \, q}. \tag{40.7}$$

It may be noticed that, in transforming (40.1) into (40.3), we did not include a term in dt to take account of a possible explicit time-dependence of the Lagrangian, since the time would there be only a parameter which would not be involved in the transformation. Analogously to formula (40.6), the partial time derivatives of L and H are related by

$$(\partial H/\partial t)_{p,q} = -(\partial L/\partial t)_{\dot{q},q}. \tag{40.8}$$

PROBLEMS

PROBLEM 1. Find the Hamiltonian for a single particle in Cartesian, cylindrical and spherical co-ordinates.

SOLUTION. In Cartesian co-ordinates x, y, z,

$$H = \frac{1}{2m}(p_x^2 + p_y^2 + p_z^2) + U(x, y, z);$$

in cylindrical co-ordinates r, ϕ, z,

$$H = \frac{1}{2m}\left(p_r^2 + \frac{p_\phi^2}{r^2} + p_z^2\right) + U(r, \phi, z);$$

in spherical co-ordinates r, θ, ϕ,

$$H = \frac{1}{2m}\left(p_r^2 + \frac{p_\theta^2}{r^2} + \frac{p_\phi^2}{r^2 \sin^2\theta}\right) + U(r, \theta, \phi).$$

PROBLEM 2. Find the Hamiltonian for a particle in a uniformly rotating frame of reference.

SOLUTION. Expressing the velocity $\mathbf{v}$ in the energy (39.11) in terms of the momentum $\mathbf{p}$ by (39.10), we have $H = p^2/2m - \mathbf{\Omega} \cdot \mathbf{r} \times \mathbf{p} + U$.

PROBLEM 3. Find the Hamiltonian for a system comprising one particle of mass M and n particles each of mass m, excluding the motion of the centre of mass (see §13, Problem).

SOLUTION. The energy E is obtained from the Lagrangian found in §13, Problem, by changing the sign of U. The generalised momenta are

$$\mathbf{p}_a = \partial L/\partial \mathbf{v}_a$$

$$= m\mathbf{v}_a - (m^2/\mu) \sum_a \mathbf{v}_a.$$

Hence

$$\sum \mathbf{p}_a = m \sum \mathbf{v}_a - (nm^2/\mu) \sum \mathbf{v}_a$$

$$= (mM/\mu) \sum \mathbf{v}_a,$$

$$\mathbf{v}_a = \mathbf{p}_a/m + (1/M)\sum \mathbf{p}_c.$$

Substitution in E gives

$$H = \frac{1}{2m} \sum_a \mathbf{p}_a^2 + \frac{1}{2M}\left(\sum_a \mathbf{p}_a\right)^2 + U.$$

§41. The Routhian

In some cases it is convenient, in changing to new variables, to replace only some, and not all, of the generalised velocities by momenta. The transformation is entirely similar to that given in §40.

To simplify the formulae, let us at first suppose that there are only two co-ordinates q and ξ, say, and transform from the variables q, ξ, $\dot{q}$, $\dot{\xi}$ to q, ξ, p, $\dot{\xi}$, where p is the generalised momentum corresponding to the co-ordinate q.

The differential of the Lagrangian $L(q, \xi, \dot{q}, \dot{\xi})$ is

$$dL = (\partial L/\partial q)\, dq + (\partial L/\partial \dot{q})\, d\dot{q} + (\partial L/\partial \xi)\, d\xi + (\partial L/\partial \dot{\xi})\, d\dot{\xi}$$
$$= \dot{p}\, dq + p\, d\dot{q} + (\partial L/\partial \xi)\, d\xi + (\partial L/\partial \dot{\xi})\, d\dot{\xi},$$

whence

$$d(L - p\dot{q}) = \dot{p}\, dq - \dot{q}\, dp + (\partial L/\partial \xi)\, d\xi + (\partial L/\partial \dot{\xi})\, d\dot{\xi}.$$

If we define the *Routhian* as

$$R(q, p, \xi, \dot{\xi}) = p\dot{q} - L, \tag{41.1}$$

in which the velocity $\dot{q}$ is expressed in terms of the momentum p by means of the equation $p = \partial L/\partial \dot{q}$, then its differential is

$$dR = -\dot{p}\, dq + \dot{q}\, dp - (\partial L/\partial \xi)\, d\xi - (\partial L/\partial \dot{\xi})\, d\dot{\xi}. \tag{41.2}$$

Hence

$$\dot{q} = \partial R/\partial p, \qquad \dot{p} = -\partial R/\partial q, \tag{41.3}$$

$$\partial L/\partial \xi = -\partial R/\partial \xi, \qquad \partial L/\partial \dot{\xi} = -\partial R/\partial \dot{\xi}. \tag{41.4}$$

Substituting these equations in the Lagrangian for the co-ordinate ξ, we have

$$\frac{d}{dt}\left(\frac{\partial R}{\partial \dot{\xi}}\right) = \frac{\partial R}{\partial \xi}. \tag{41.5}$$

Thus the Routhian is a Hamiltonian with respect to the co-ordinate q (equations (41.3)) and a Lagrangian with respect to the co-ordinate ξ (equation (41.5)).

According to the general definition the energy of the system is

$$E = \dot{q}\, \partial L/\partial \dot{q} + \dot{\xi}\, \partial L/\partial \dot{\xi} - L = p\dot{q} + \dot{\xi}\, \partial L/\partial \dot{\xi} - L.$$

In terms of the Routhian it is

$$E = R - \dot{\xi}\, \partial R/\partial \dot{\xi}, \tag{41.6}$$

as we find by substituting (41.1) and (41.4).

The generalisation of the above formulae to the case of several co-ordinates q and ξ is evident.

The use of the Routhian may be convenient, in particular, when some of the co-ordinates are cyclic. If the co-ordinates q are cyclic, they do not appear in the Lagrangian, nor therefore in the Routhian, so that the latter is a function of p, ξ and $\dot{\xi}$. The momenta p corresponding to cyclic co-ordinates are constant, as follows also from the second equation (41.3), which in this sense contains no new information. When the momenta p are replaced by their given constant values, equations (41.5) $(d/dt)\, \partial R(p, \xi, \dot{\xi})/\partial \dot{\xi} = \partial R(p, \xi, \dot{\xi})/\partial \xi$ become equations containing only the co-ordinates ξ, so that the cyclic co-ordinates are entirely eliminated. If these equations are solved for the functions $\xi(t)$, substitution of the latter on the right-hand sides of the equations $\dot{q} = \partial R(p, \xi, \dot{\xi})/\partial p$ gives the functions $q(t)$ by direct integration.

PROBLEM

Find the Routhian for a symmetrical top in an external field $U(\phi, \theta)$, eliminating the cyclic co-ordinate ψ (where ψ, ϕ, θ are Eulerian angles).

SOLUTION. The Lagrangian is $L = \frac{1}{2}I'_1(\dot\theta^2+\dot\phi^2\sin^2\theta)+\frac{1}{2}I_3(\dot\psi+\dot\phi\cos\theta)^2-U(\phi,\ \theta)$; see §35, Problem 1. The Routhian is

$$R = p_\psi\dot\psi-L = \frac{p_\psi^2}{2I_3}-p_\psi\dot\phi\cos\theta-\frac{1}{2}I'_1(\dot\theta^2+\dot\phi^2\sin^2\theta)+U(\phi,\ \theta);$$

the first term is a constant and may be omitted.

§42. Poisson brackets

Let $f(p,\ q,\ t)$ be some function of co-ordinates, momenta and time. Its total time derivative is

$$\frac{df}{dt} = \frac{\partial f}{\partial t}+\sum_k\left(\frac{\partial f}{\partial q_k}\dot q_k+\frac{\partial f}{\partial p_k}\dot p_k\right).$$

Substitution of the values of $\dot q_k$ and $\dot p_k$ given by Hamilton's equations (40.4) leads to the expression

$$df/dt = \partial f/\partial t+[H, f], \tag{42.1}$$

where

$$[H, f] \equiv \sum_k\left(\frac{\partial H}{\partial p_k}\frac{\partial f}{\partial q_k}-\frac{\partial H}{\partial q_k}\frac{\partial f}{\partial p_k}\right). \tag{42.2}$$

This expression is called the *Poisson bracket* of the quantities H and f.

Those functions of the dynamical variables which remain constant during the motion of the system are, as we know, called *integrals of the motion*. We see from (42.1) that the condition for the quantity f to be an integral of the motion ($df/dt = 0$) can be written

$$\partial f/\partial t+[H, f] = 0. \tag{42.3}$$

If the integral of the motion is not explicitly dependent on the time, then

$$[H, f] = 0, \tag{42.4}$$

i.e. the Poisson bracket of the integral and the Hamiltonian must be zero.

For any two quantities f and g, the Poisson bracket is defined analogously to (42.2):

$$[f, g] \equiv \sum_k\left(\frac{\partial f}{\partial p_k}\frac{\partial g}{\partial q_k}-\frac{\partial f}{\partial q_k}\frac{\partial g}{\partial p_k}\right). \tag{42.5}$$

The Poisson bracket has the following properties, which are easily derived from its definition.

If the two functions are interchanged, the bracket changes sign; if one of the functions is a constant c, the bracket is zero:

$$[f, g] = -[g, f], \tag{42.6}$$

$$[f, c] = 0. \tag{42.7}$$

Also

$$[f_1+f_2, g] = [f_1, g]+[f_2, g], \tag{42.8}$$

$$[f_1f_2, g] = f_1[f_2, g]+f_2[f_1, g]. \tag{42.9}$$

Taking the partial derivative of (42.5) with respect to time, we obtain

$$\frac{\partial}{\partial t}[f, g] = \left[\frac{\partial f}{\partial t}, g\right]+\left[f, \frac{\partial g}{\partial t}\right]. \tag{42.10}$$

If one of the functions f and g is one of the momenta or co-ordinates, the Poisson bracket reduces to a partial derivative:

$$[f, q_k] = \partial f/\partial p_k, \tag{42.11}$$

$$[f, p_k] = -\partial f/\partial q_k. \tag{42.12}$$

Formula (42.11), for example, may be obtained by putting $g = q_k$ in (42.5); the sum reduces to a single term, since $\partial q_k/\partial q_l = \delta_{kl}$ and $\partial q_k/\partial p_l = 0$. Putting in (42.11) and (42.12) the function f equal to q_i and p_i we have, in particular,

$$[q_i, q_k] = 0, \quad [p_i, p_k] = 0, \quad [p_i, q_k] = \delta_{ik}. \tag{42.13}$$

The relation

$$[f, [g, h]] + [g, [h, f]] + [h, [f, g]] = 0, \tag{42.14}$$

known as *Jacobi's identity*, holds between the Poisson brackets formed from three functions f, g and h. To prove it, we first note the following result. According to the definition (42.5), the Poisson bracket $[f, g]$ is a bilinear homogeneous function of the first derivatives of f and g. Hence the bracket $[h, [f, g]]$, for example, is a linear homogeneous function of the second derivatives of f and g. The left-hand side of equation (42.14) is therefore a linear homogeneous function of the second derivatives of all three functions f, g and h. Let us collect the terms involving the second derivatives of f. The first bracket contains no such terms, since it involves only the first derivatives of f. The sum of the second and third brackets may be symbolically written in terms of the linear differential operators D_1 and D_2, defined by $D_1(\phi) = [g, \phi]$, $D_2(\phi) = [h, \phi]$. Then

$$[g, [h, f]] + [h, [f, g]] = [g, [h, f]] - [h, [g, f]]$$
$$= D_1[D_2(f)] - D_2[D_1(f)]$$
$$= (D_1 D_2 - D_2 D_1)f.$$

It is easy to see that this combination of linear differential operators cannot involve the second derivatives of f. The general form of the linear differential operators is

$$D_1 = \sum_k \xi_k \frac{\partial}{\partial x_k}, \quad D_2 = \sum_k \eta_k \frac{\partial}{\partial x_k},$$

where ξ_k and η_k are arbitrary functions of the variables $x_1, x_2, \dots$. Then

$$D_1 D_2 = \sum_{k,l} \xi_k \eta_l \frac{\partial^2}{\partial x_k \partial x_l} + \sum_{k,l} \xi_k \frac{\partial \eta_l}{\partial x_k} \frac{\partial}{\partial x_l},$$

$$D_2 D_1 = \sum_{k,l} \eta_k \xi_l \frac{\partial^2}{\partial x_k \partial x_l} + \sum_{k,l} \eta_k \frac{\partial \xi_l}{\partial x_k} \frac{\partial}{\partial x_l},$$

and the difference of these,

$$D_1D_2 - D_2D_1 = \sum_{k,l}\left(\xi_k\frac{\partial\eta_l}{\partial x_k} - \eta_k\frac{\partial\xi_l}{\partial x_k}\right)\frac{\partial}{\partial x_l},$$

is again an operator involving only single differentiations. Thus the terms in the second derivatives of f on the left-hand side of equation (42.14) cancel and, since the same is of course true of g and h, the whole expression is identically zero.

An important property of the Poisson bracket is that, if f and g are two integrals of the motion, their Poisson bracket is likewise an integral of the motion:

$$[f, g] = \text{constant.} \tag{42.15}$$

This is *Poisson's theorem*. The proof is very simple if f and g do not depend explicitly on the time. Putting $h = H$ in Jacobi's identity, we obtain

$$[H, [f, g]] + [f, [g, H]] + [g, [H, f]] = 0.$$

Hence, if $[H, g] = 0$ and $[H, f] = 0$, then $[H, [f, g]] = 0$, which is the required result.

If the integrals f and g of the motion are explicitly time-dependent, we put, from (42.1),

$$\frac{d}{dt}[f, g] = \frac{\partial}{\partial t}[f, g] + [H, [f, g]].$$

Using formula (42.10) and expressing the bracket $[H, [f, g]]$ in terms of two others by means of Jacobi's identity, we find

$$\frac{d}{dt}[f, g] = \left[\frac{\partial f}{\partial t}, g\right] + \left[f, \frac{\partial g}{\partial t}\right] - [f, [g, H]] - [g, [H, f]]$$

$$= \left[\frac{\partial f}{\partial t} + [H, f], g\right] + \left[f, \frac{\partial g}{\partial t} + [H, g]\right]$$

$$= \left[\frac{df}{dt}, g\right] + \left[f, \frac{dg}{dt}\right], \tag{42.16}$$

which evidently proves Poisson's theorem.

Of course, Poisson's theorem does not always supply further integrals of the motion, since there are only $2s - 1$ of these (s being the number of degrees of freedom). In some cases the result is trivial, the Poisson bracket being a constant. In other cases the integral obtained is simply a function of the original integrals f and g. If neither of these two possibilities occurs, however, then the Poisson bracket is a further integral of the motion.

PROBLEMS

PROBLEM 1. Determine the Poisson brackets formed from the Cartesian components of the momentum $\mathbf{p}$ and the angular momentum $\mathbf{M} = \mathbf{r}\times\mathbf{p}$ of a particle.

SOLUTION. Formula (42.12) gives $[M_x, p_y] = -\partial M_x/\partial y = -\partial(yp_z - zp_y)/\partial y = -p_z$, and similarly $[M_x, p_x] = 0$, $[M_x, p_z] = p_y$. The remaining brackets are obtained by cyclically permuting the suffixes x, y, z.

PROBLEM 2. Determine the Poisson brackets formed from the components of **M**.

SOLUTION. A direct calculation from formula (42.5) gives $[M_x, M_y] = -M_z$, $[M_y, M_z]$ $= -M_x$, $[M_z, M_x] = -M_y$.

Since the momenta and co-ordinates of different particles are mutually independent variables, it is easy to see that the formulae derived in Problems 1 and 2 are valid also for the total momentum and angular momentum of any system of particles.

PROBLEM 3. Show that $[\phi, M_z] = 0$, where ϕ is any function, spherically symmetrical about the origin, of the co-ordinates and momentum of a particle.

SOLUTION. Such a function ϕ can depend on the components of the vectors **r** and **p** only through the combinations $\mathbf{r}^2$, $\mathbf{p}^2$, $\mathbf{r} \cdot \mathbf{p}$. Hence

$$\frac{\partial \phi}{\partial \mathbf{r}} = \frac{\partial \phi}{\partial (\mathbf{r}^2)} \cdot 2\mathbf{r} + \frac{\partial \phi}{\partial (\mathbf{p} \cdot \mathbf{r})} \cdot \mathbf{p},$$

and similarly for $\partial \phi / \partial \mathbf{p}$. The required relation may be verified by direct calculation from formula (42.5), using these formulae for the partial derivatives.

PROBLEM 4. Show that $[\mathbf{f}, M_z] = \mathbf{f} \times \mathbf{n}$, where **f** is a vector function of the co-ordinates and momentum of a particle, and **n** is a unit vector parallel to the z-axis.

SOLUTION. An arbitrary vector $\mathbf{f}(\mathbf{r}, \mathbf{p})$ may be written as $\mathbf{f} = \mathbf{r}\phi_1 + \mathbf{p}\phi_2 + \mathbf{r} \times \mathbf{p}\phi_3$, where ϕ_1, ϕ_2, ϕ_3 are scalar functions. The required relation may be verified by direct calculation from formulae (42.9), (42.11), (42.12) and the formula of Problem 3.

§43. The action as a function of the co-ordinates

In formulating the principle of least action, we have considered the integral

$$S = \int_{t_1}^{t_2} L \, dt, \tag{43.1}$$

taken along a path between two given positions $q^{(1)}$ and $q^{(2)}$ which the system occupies at given instants t_1 and t_2. In varying the action, we compared the values of this integral for neighbouring paths with the same values of $q(t_1)$ and $q(t_2)$. Only one of these paths corresponds to the actual motion, namely the path for which the integral S has its minimum value.

Let us now consider another aspect of the concept of action, regarding S as a quantity characterising the motion along the actual path, and compare the values of S for paths having a common beginning at $q(t_1) = q^{(1)}$, but passing through different points at time t_2. In other words, we consider the action integral for the true path as a function of the co-ordinates at the upper limit of integration.

The change in the action from one path to a neighbouring path is given (if there is one degree of freedom) by the expression (2.5):

$$\delta S = \left[\frac{\partial L}{\partial \dot{q}} \delta q \right]_{t_1}^{t_2} + \int_{t_1}^{t_2} \left(\frac{\partial L}{\partial q} - \frac{d}{dt} \frac{\partial L}{\partial \dot{q}} \right) \delta q \, dt.$$

Since the paths of actual motion satisfy Lagrange's equations, the integral in δS is zero. In the first term we put $\delta q(t_1) = 0$, and denote the value of

$\delta q(t_2)$ by δq simply. Replacing $\partial L/\partial \dot{q}$ by p, we have finally $\delta S = p\delta q$ or, in the general case of any number of degrees of freedom,

$$\delta S = \sum_i p_i \delta q_i. \qquad (43.2)$$

From this relation it follows that the partial derivatives of the action with respect to the co-ordinates are equal to the corresponding momenta:

$$\partial S/\partial q_i = p_i. \qquad (43.3)$$

The action may similarly be regarded as an explicit function of time, by considering paths starting at a given instant t_1 and at a given point $q^{(1)}$, and ending at a given point $q^{(2)}$ at various times $t_2 = t$. The partial derivative $\partial S/\partial t$ thus obtained may be found by an appropriate variation of the integral. It is simpler, however, to use formula (43.3), proceeding as follows.

From the definition of the action, its total time derivative along the path is

$$dS/dt = L. \qquad (43.4)$$

Next, regarding S as a function of co-ordinates and time, in the sense described above, and using formula (43.3), we have

$$\frac{dS}{dt} = \frac{\partial S}{\partial t} + \sum_i \frac{\partial S}{\partial q_i} \dot{q}_i = \frac{\partial S}{\partial t} + \sum_i p_i \dot{q}_i.$$

A comparison gives $\partial S/\partial t = L - \sum p_i \dot{q}_i$ or

$$\partial S/\partial t = -H. \qquad (43.5)$$

Formulae (43.3) and (43.5) may be represented by the expression

$$dS = \sum_i p_i\, dq_i - H\, dt \qquad (43.6)$$

for the total differential of the action as a function of co-ordinates and time at the upper limit of integration in (43.1). Let us now suppose that the co-ordinates (and time) at the beginning of the motion, as well as at the end, are variable. It is evident that the corresponding change in S will be given by the difference of the expressions (43.6) for the beginning and end of the path, i.e.

$$dS = \sum p_i^{(2)}\, dq_i^{(2)} - H^{(2)}\, dt^{(2)} - \sum p_i^{(1)}\, dq_i^{(1)} + H^{(1)}\, dt^{(1)}. \qquad (43.7)$$

This relation shows that, whatever the external forces on the system during its motion, its final state cannot be an arbitrary function of its initial state; only those motions are possible for which the expression on the right-hand side of equation (43.7) is a perfect differential. Thus the existence of the principle of least action, quite apart from any particular form of the Lagrangian, imposes certain restrictions on the range of possible motions. In particular, it is possible to derive a number of general properties, independent of the external fields, for beams of particles diverging from given points in space. The study of these properties forms a part of the subject of *geometrical optics.*†

It is of interest to note that Hamilton's equations can be formally derived from the condition of minimum action in the form

$$S = \int \left(\sum_i p_i \, dq_i - H \, dt \right), \tag{43.8}$$

which follows from (43.6), if the co-ordinates and momenta are varied independently. Again assuming for simplicity that there is only one co-ordinate and momentum, we write the variation of the action as

$$\delta S = \int [\delta p \, dq + p \, d\delta q - (\partial H/\partial q)\delta q \, dt - (\partial H/\partial p)\delta p \, dt].$$

An integration by parts in the second term gives

$$\delta S = \int \delta p \{dq - (\partial H/\partial p) \, dt\} + [p\delta q] - \int \delta q \{dp + (\partial H/\partial q) \, dt\}.$$

At the limits of integration we must put $\delta q = 0$, so that the integrated term is zero. The remaining expression can be zero only if the two integrands vanish separately, since the variations δp and δq are independent and arbitrary: $dq = (\partial H/\partial p) \, dt$, $dp = -(\partial H/\partial q) \, dt$, which, after division by dt, are Hamilton's equations.

§44. Maupertuis' principle

The motion of a mechanical system is entirely determined by the principle of least action: by solving the equations of motion which follow from that principle, we can find both the form of the path and the position on the path as a function of time.

If the problem is the more restricted one of determining only the path, without reference to time, a simplified form of the principle of least action may be used. We assume that the Lagrangian, and therefore the Hamiltonian, do not involve the time explicitly, so that the energy of the system is conserved: $H(p, q) = E =$ constant. According to the principle of least action, the variation of the action, for given initial and final co-ordinates and times (t_0 and t, say), is zero. If, however, we allow a variation of the final time t, the initial and final co-ordinates remaining fixed, we have (cf.(43.7))

$$\delta S = -H\delta t. \tag{44.1}$$

We now compare, not all virtual motions of the system, but only those which satisfy the law of conservation of energy. For such paths we can replace H in (44.1) by a constant E, which gives

$$\delta S + E\delta t = 0. \tag{44.2}$$

† See *The Classical Theory of Fields*, Chapter 7, Pergamon Press, Oxford 1975.

Writing the action in the form (43.8) and again replacing H by E, we have

$$S = \int \sum_i p_i \, dq_i - E(t - t_0). \tag{44.3}$$

The first term in this expression,

$$S_0 = \int \sum_i p_i \, dq_i, \tag{44.4}$$

is sometimes called the *abbreviated action*.

Substituting (44.3) in (44.2), we find that

$$\delta S_0 = 0. \tag{44.5}$$

Thus the abbreviated action has a minimum with respect to all paths which satisfy the law of conservation of energy and pass through the final point at any instant. In order to use such a variational principle, the momenta (and so the whole integrand in (44.4)) must be expressed in terms of the co-ordinates q and their differentials dq. To do this, we use the definition of momentum:

$$p_i = \frac{\partial}{\partial \dot{q}_i} L\left(q, \frac{dq}{dt} \right) \tag{44.6}$$

and the law of conservation of energy:

$$E\left(q, \frac{dq}{dt} \right) = E. \tag{44.7}$$

Expressing the differential dt in terms of the co-ordinates q and their differentials dq by means of (44.7) and substituting in (44.6), we have the momenta in terms of q and dq, with the energy E as a parameter. The variational principle so obtained determines the path of the system, and is usually called *Maupertuis' principle*, although its precise formulation is due to EULER and LAGRANGE.

The above calculations may be carried out explicitly when the Lagrangian takes its usual form (5.5) as the difference of the kinetic and potential energies:

$$L = \tfrac{1}{2} \sum_{i,k} a_{ik}(q) \dot{q}_i \dot{q}_k - U(q).$$

The momenta are

$$p_i = \partial L / \partial \dot{q}_i = \sum_k a_{ik}(q) \dot{q}_k,$$

and the energy is

$$E = \tfrac{1}{2} \sum_{i,k} a_{ik}(q) \dot{q}_i \dot{q}_k + U(q).$$

The last equation gives

$$dt = \sqrt{\left[\sum a_{ik} \, dq_i \, dq_k / 2(E - U) \right]}; \tag{44.8}$$

substituting this in

$$\sum_i p_i \, dq_i = \sum_{i,k} a_{ik} \frac{dq_k}{dt} \, dq_i,$$

we find the abbreviated action:

$$S_0 = \int \sqrt{[\, 2(E-U) \sum_{i,k} a_{ik} \, dq_i \, dq_k \,]}. \tag{44.9}$$

In particular, for a single particle the kinetic energy is $T = \frac{1}{2} m(dl/dt)^2$, where m is the mass of the particle and dl an element of its path; the variational principle which determines the path is

$$\delta \int \sqrt{[2m(E-U)]} \, dl = 0, \tag{44.10}$$

where the integral is taken between two given points in space. This form is due to JACOBI.

In free motion of the particle, $U = 0$, and (44.10) gives the trivial result $\delta \int dl = 0$, i.e. the particle moves along the shortest path between the two given points, i.e. in a straight line.

Let us return now to the expression (44.3) for the action and vary it with respect to the parameter E. We have

$$\delta S = \frac{\partial S_0}{\partial E} \, \delta E - (t - t_0)\delta E - E\delta t;$$

substituting in (44.2), we obtain

$$\partial S_0 / \partial E = t - t_0. \tag{44.11}$$

When the abbreviated action has the form (44.9), this gives

$$\int \sqrt{[\sum a_{ik} \, dq_i \, dq_k / 2(E-U)]} = t - t_0, \tag{44.12}$$

which is just the integral of equation (44.8). Together with the equation of the path, it entirely determines the motion.

PROBLEM

Derive the differential equation of the path from the variational principle (44.10).

SOLUTION. Effecting the variation, we have

$$\delta \int \sqrt{(E-U)} \, dl = -\int \left\{ \frac{\partial U}{\partial \mathbf{r}} \cdot \frac{\delta \mathbf{r}}{2\sqrt{(E-U)}} \, dl - \sqrt{(E-U)} \frac{d\mathbf{r}}{dl} \cdot d\delta \mathbf{r} \right\}.$$

In the second term we have used the fact that $dl^2 = d\mathbf{r}^2$ and therefore $dl \, d\delta l = d\mathbf{r} \cdot d\delta \mathbf{r}$. Integrating this term by parts and then equating to zero the coefficient of $\delta \mathbf{r}$ in the integrand, we obtain the differential equation of the path:

$$2\sqrt{(E-U)} \frac{d}{dl} \left[\sqrt{(E-U)} \frac{d\mathbf{r}}{dl} \right] = -\partial U / \partial \mathbf{r}.$$

Expanding the derivative on the left-hand side and putting the force $\mathbf{F} = -\partial U/\partial \mathbf{r}$ gives

$$d^2\mathbf{r}/dl^2 = [\mathbf{F}-(\mathbf{F}\cdot\mathbf{t})\mathbf{t}]/2(E-U),$$

where $\mathbf{t} = d\mathbf{r}/dl$ is a unit vector tangential to the path. The difference $\mathbf{F}-(\mathbf{F}\cdot\mathbf{t})\mathbf{t}$ is the component $\mathbf{F}_n$ of the force normal to the path. The derivative $d^2\mathbf{r}/dl^2 = d\mathbf{t}/dl$ is known from differential geometry to be $\mathbf{n}/R$, where R is the radius of curvature of the path and $\mathbf{n}$ the unit vector along the principal normal. Replacing $E-U$ by $\frac{1}{2}mv^2$, we have $(mv^2/R)\mathbf{n} = \mathbf{F}_n$, in agreement with the familar expression for the normal acceleration in motion in a curved path.

§45. Canonical transformations

The choice of the generalised co-ordinates q is subject to no restriction; they may be any s quantities which uniquely define the position of the system in space. The formal appearance of Lagrange's equations (2.6) does not depend on this choice, and in that sense the equations may be said to be invariant with respect to a transformation from the co-ordinates $q_1, q_2, ...$ to any other independent quantities $Q_1, Q_2, ...$. The new co-ordinates Q are functions of q, and we shall assume that they may explicitly depend on the time, i.e. that the transformation is of the form

$$Q_i = Q_i(q, t) \tag{45.1}$$

(sometimes called a *point transformation*).

Since Lagrange's equations are unchanged by the transformation (45.1), Hamilton's equations (40.4) are also unchanged. The latter equations, however, in fact allow a much wider range of transformations. This is, of course, because in the Hamiltonian treatment the momenta p are variables independent of and on an equal footing with the co-ordinates q. Hence the transformation may be extended to include all the $2s$ independent variables p and q:

$$Q_i = Q_i(p, q, t), \qquad P_i = P_i(p, q, t). \tag{45.2}$$

This enlargement of the class of possible transformations is one of the important advantages of the Hamiltonian treatment.

The equations of motion do not, however, retain their canonical form under all transformations of the form (45.2). Let us derive the conditions which must be satisfied if the equations of motion in the new variables P, Q are to be of the form

$$\dot{Q}_i = \partial H'/\partial P_i, \qquad \dot{P}_i = -\partial H'/\partial Q_i \tag{45.3}$$

with some Hamiltonian $H'(P, Q)$. When this happens the transformation is said to be *canonical*.

The formulae for canonical transformations can be obtained as follows. It has been shown at the end of §43 that Hamilton's equations can be derived from the principle of least action in the form

$$\delta\int \left(\sum_i p_i \, dq_i - H \, dt\right) = 0, \tag{45.4}$$

in which the variation is applied to all the co-ordinates and momenta independently. If the new variables P and Q also satisfy Hamilton's equations, the principle of least action

$$\delta\int(\sum_i P_i\,dQ_i - H'\,dt) = 0 \qquad (45.5)$$

must hold. The two forms (45.4) and (45.5) are equivalent only if their integrands are the same apart from the total differential of some function F of co-ordinates, momenta and time.† The difference between the two integrals is then a constant, namely the difference of the values of F at the limits of integration, which does not affect the variation. Thus we must have

$$\sum p_i\,dq_i - H\,dt = \sum P_i\,dQ_i - H'\,dt + dF.$$

Each canonical transformation is characterised by a particular function F, called the *generating function* of the transformation.

Writing this relation as

$$dF = \sum p_i\,dq_i - \sum P_i\,dQ_i + (H' - H)\,dt, \qquad (45.6)$$

we see that

$$p_i = \partial F/\partial q_i, \qquad P_i = -\partial F/\partial Q_i, \qquad H' = H + \partial F/\partial t; \qquad (45.7)$$

here it is assumed that the generating function is given as a function of the old and new co-ordinates and the time: $F = F(q, Q, t)$. When F is known, formulae (45.7) give the relation between p, q and P, Q as well as the new Hamiltonian.

It may be convenient to express the generating function not in terms of the variables q and Q but in terms of the old co-ordinates q and the new momenta P. To derive the formulae for canonical transformations in this case, we must effect the appropriate Legendre's transformation in (45.6), rewriting it as

$$d(F + \sum P_i Q_i) = \sum p_i\,dq_i + \sum Q_i\,dP_i + (H' - H)\,dt.$$

The argument of the differential on the left-hand side, expressed in terms of the variables q and P, is a new generating function $\Phi(q, P, t)$, say. Then‡

$$p_i = \partial\Phi/\partial q_i, \qquad Q_i = \partial\Phi/\partial P_i, \qquad H' = H + \partial\Phi/\partial t. \qquad (45.8)$$

We can similarly obtain the formulae for canonical transformations involving generating functions which depend on the variables p and Q, or p and P.

---

† We do not consider such trivial transformations as $P_i = ap_i$, $Q_i = q_i$, $H' = aH$, with a an arbitrary constant, whereby the integrands in (45.4) and (45.5) differ only by a constant factor.

‡ If the generating function is $\Phi = \Sigma f_i(q, t)P_i$, where the f_i are arbitrary functions, we obtain a transformation in which the new co-ordinates are $Q_i = f_i(q, t)$, i.e. are expressed in terms of the old co-ordinates only (and not the momenta). This is a point transformation, and is of course a particular canonical transformation.

The relation between the two Hamiltonians is always of the same form: the difference $H' - H$ is the partial derivative of the generating function with respect to time. In particular, if the generating function is independent of time, then $H' = H$, i.e. the new Hamiltonian is obtained by simply substituting for p, q in H their values in terms of the new variables P, Q.

The wide range of the canonical transformations in the Hamiltonian treatment deprives the generalised co-ordinates and momenta of a considerable part of their original meaning. Since the transformations (45.2) relate each of the quantities P, Q to both the co-ordinates q and the momenta p, the variables Q are no longer purely spatial co-ordinates, and the distinction between Q and P becomes essentially one of nomenclature. This is very clearly seen, for example, from the transformation† $Q_i = p_i$, $P_i = -q_i$, which obviously does not affect the canonical form of the equations and amounts simply to calling the co-ordinates momenta and *vice versa*.

On account of this arbitrariness of nomenclature, the variables p and q in the Hamiltonian treatment are often called simply *canonically conjugate* quantities. The conditions relating such quantities can be expressed in terms of Poisson brackets. To do this, we shall first prove a general theorem on the invariance of Poisson brackets with respect to canonical transformations.

Let $[f, g]_{p,q}$ be the Poisson bracket, for two quantities f and g, in which the differentiation is with respect to the variables p and q, and $[f, g]_{P,Q}$ that in which the differentiation is with respect to P and Q. Then

$$[f, g]_{p,q} = [f, g]_{P,Q}. \qquad (45.9)$$

The truth of this statement can be seen by direct calculation, using the formulae of the canonical transformation. It can also be demonstrated by the following argument.

First of all, it may be noticed that the time appears as a parameter in the canonical transformations (45.7) and (45.8). It is therefore sufficient to prove (45.9) for quantities which do not depend explicitly on time. Let us now formally regard g as the Hamiltonian of some fictitious system. Then, by formula (42.1), $[f, g]_{p,q} = -df/dt$. The derivative df/dt can depend only on the properties of the motion of the fictitious system, and not on the particular choice of variables. Hence the Poisson bracket $[f, g]$ is unaltered by the passage from one set of canonical variables to another.

Formulae (42.13) and (45.9) give

$$[Q_i, Q_k]_{p,q} = 0, \quad [P_i, P_k]_{p,q} = 0, \quad [P_i, Q_k]_{p,q} = \delta_{ik}. \qquad (45.10)$$

These are the conditions, written in terms of Poisson brackets, which must be satisfied by the new variables if the transformation p, $q \to P$, Q is canonical.

It is of interest to observe that the change in the quantities p, q during the motion may itself be regarded as a series of canonical transformations. The meaning of this statement is as follows. Let q_t, p_t be the values of the canonical

---

† Whose generating function is $F = \Sigma q_i Q_i$.

6*

variables at time t, and $q_{t+\tau}$, $p_{t+\tau}$ their values at another time $t+\tau$. The latter are some functions of the former (and involve τ as a parameter):

$$q_{t+\tau} = q(q_t, p_t, \tau), \qquad p_{t+\tau} = p(q_t, p_t, \tau).$$

If these formulae are regarded as a transformation from the variables q_t, p_t to $q_{t+\tau}$, $p_{t+\tau}$, then this transformation is canonical. This is evident from the expression $dS = \Sigma(p_{t+\tau}dq_{t+\tau} - p_t dq_t)$ for the differential of the action $S(q_{t+\tau}, q_t)$ taken along the true path, passing through the points q_t and $q_{t+\tau}$ at given times t and $t + \tau$ (cf. (43.7)). A comparison of this formula with (45.6) shows that $-S$ is the generating function of the transformation.

§46. Liouville's theorem

For the geometrical interpretation of mechanical phenomena, use is often made of *phase space*. This is a space of $2s$ dimensions, whose co-ordinate axes correspond to the s generalised co-ordinates and s momenta of the system concerned. Each point in phase space corresponds to a definite state of the system. When the system moves, the point representing it describes a curve called the *phase path*.

The product of differentials $d\Gamma = dq_1 \ldots dq_s dp_1 \ldots dp_s$ may be regarded as an element of volume in phase space. Let us now consider the integral $\int d\Gamma$ taken over some region of phase space, and representing the volume of that region. We shall show that this integral is invariant with respect to canonical transformations; that is, if the variables p, q are replaced by P, Q by a canonical transformation, then the volumes of the corresponding regions of the spaces of p, q and P, Q are equal:

$$\int \ldots \int dq_1 \ldots dq_s \, dp_1 \ldots dp_s = \int \ldots \int dQ_1 \ldots dQ_s \, dP_1 \ldots dP_s. \qquad (46.1)$$

The transformation of variables in a multiple integral is effected by the formula $\int \ldots \int dQ_1 \ldots dQ_s \, dP_1 \ldots dP_s = \int \ldots \int D dq_1 \ldots dq_s \, dp_1 \ldots dp_s,$ where

$$D = \frac{\partial(Q_1, \ldots, Q_s, P_1, \ldots, P_s)}{\partial(q_1, \ldots, q_s, p_1, \ldots, p_s)} \qquad (46.2)$$

is the Jacobian of the transformation. The proof of (46.1) therefore amounts to proving that the Jacobian of every canonical transformation is unity:

$$D = 1. \qquad (46.3)$$

We shall use a well-known property of Jacobians whereby they can be treated somewhat like fractions. "Dividing numerator and denominator" by $\partial(q_1, \ldots, q_s, P_1, \ldots, P_s)$, we obtain

$$D = \frac{\partial(Q_1, \ldots, Q_s, P_1, \ldots, P_s)}{\partial(q_1, \ldots, q_s, P_1, \ldots, P_s)} \bigg/ \frac{\partial(q_1, \ldots, q_s, p_1, \ldots, p_s)}{\partial(q_1, \ldots, q_s, P_1, \ldots, P_s)}.$$

Another property of Jacobians is that, when the same quantities appear in both the partial differentials, the Jacobian reduces to one in fewer variables, in which these repeated quantities are regarded as constant in carrying out the differentiations. Hence

$$D = \left\{ \frac{\partial(Q_1, \ldots, Q_s)}{\partial(q_1, \ldots, q_s)} \right\}_{P=\text{constant}} \bigg/ \left\{ \frac{\partial(p_1, \ldots, p_s)}{\partial(P_1, \ldots, P_s)} \right\}_{q=\text{constant}}. \quad (46.4)$$

The Jacobian in the numerator is, by definition, a determinant of order s whose element in the ith row and kth column is $\partial Q_i/\partial q_k$. Representing the canonical transformation in terms of the generating function $\Phi(q, P)$ as in (45.8), we have $\partial Q_i/\partial q_k = \partial^2\Phi/\partial q_k\partial P_i$. In the same way we find that the ik-element of the determinant in the denominator of (46.4) is $\partial^2\Phi/\partial q_i\partial P_k$. This means that the two determinants differ only by the interchange of rows and columns; they are therefore equal, so that the ratio (46.4) is equal to unity. This completes the proof.

Let us now suppose that each point in the region of phase space considered moves in the course of time in accordance with the equations of motion of the mechanical system. The region as a whole therefore moves also, but its volume remains unchanged:

$$\int d\Gamma = \text{constant}. \quad (46.5)$$

This result, known as *Liouville's theorem*, follows at once from the invariance of the volume in phase space under canonical transformations and from the fact that the change in p and q during the motion may, as we showed at the end of §45, be regarded as a canonical transformation.

In an entirely similar manner the integrals

$$\iint \sum_i dq_i\, dp_i, \qquad \iiiint \sum_{i\neq k} dq_i\, dp_i\, dq_k\, dp_k, \ldots,$$

in which the integration is over manifolds of two, four, etc. dimensions in phase space, may be shown to be invariant.

§47. The Hamilton–Jacobi equation

In §43 the action has been considered as a function of co-ordinates and time, and it has been shown that the partial derivative with respect to time of this function $S(q, t)$ is related to the Hamiltonian by

$$\partial S/\partial t + H(q, p, t) = 0,$$

and its partial derivatives with respect to the co-ordinates are the momenta. Accordingly replacing the momenta p in the Hamiltonian by the derivatives $\partial S/\partial q$, we have the equation

$$\frac{\partial S}{\partial t} + H\left(q_1, \ldots, q_s; \frac{\partial S}{\partial q_1}, \ldots, \frac{\partial S}{\partial q_s}; t\right) = 0 \quad (47.1)$$

which must be satisfied by the function $S(q, t)$. This first-order partial differential equation is called the *Hamilton–Jacobi equation*.

Like Lagrange's equations and the canonical equations, the Hamilton–Jacobi equation is the basis of a general method of integrating the equations of motion.

Before describing this method, we should recall the fact that every first-order partial differential equation has a solution depending on an arbitrary function; such a solution is called the *general integral* of the equation. In mechanical applications, the general integral of the Hamilton–Jacobi equation is less important than a *complete integral*, which contains as many independent arbitrary constants as there are independent variables.

The independent variables in the Hamilton–Jacobi equation are the time and the co-ordinates. For a system with s degrees of freedom, therefore, a complete integral of this equation must contain $s+1$ arbitrary constants. Since the function S enters the equation only through its derivatives, one of these constants is additive, so that a complete integral of the Hamilton–Jacobi equation is

$$S = f(t, q_1, ..., q_s; \alpha_1, ..., \alpha_s) + A, \qquad (47.2)$$

where $\alpha_1, ..., \alpha_s$ and A are arbitrary constants.†

Let us now ascertain the relation between a complete integral of the Hamilton–Jacobi equation and the solution of the equations of motion which is of interest. To do this, we effect a canonical transformation from the variables q, p to new variables, taking the function $f(t, q; \alpha)$ as the generating function, and the quantities $\alpha_1, \alpha_2, ..., \alpha_s$ as the new momenta. Let the new co-ordinates be $\beta_1, \beta_2, ..., \beta_s$. Since the generating function depends on the old co-ordinates and the new momenta, we use formulae (45.8): $p_i = \partial f/\partial q_i$, $\beta_i = \partial f/\partial \alpha_i$, $H' = H + \partial f/\partial t$. But since the function f satisfies the Hamilton–Jacobi equation, we see that the new Hamiltonian is zero: $H' = H + \partial f/\partial t = H + \partial S/\partial t = 0$. Hence the canonical equations in the new variables are $\dot{\alpha}_i = 0$, $\dot{\beta}_i = 0$, whence

$$\alpha_i = \text{constant}, \qquad \beta_i = \text{constant}. \qquad (47.3)$$

By means of the s equations $\partial f/\partial \alpha_i = \beta_i$, the s co-ordinates q can be expressed in terms of the time and the $2s$ constants α and β. This gives the general integral of the equations of motion.

---

† Although the general integral of the Hamilton–Jacobi equation is not needed here, we may show how it can be found from a complete integral. To do this, we regard A as an arbitrary function of the remaining constants: $S = f(t, q_1, ..., q_s; \alpha_1, ..., \alpha_s) + A(\alpha_1, ..., \alpha_s)$. Replacing the α_i by functions of co-ordinates and time given by the s conditions $\partial S/\partial \alpha_i = 0$, we obtain the general integral in terms of the arbitrary function $A(\alpha_1, ..., \alpha_s)$. For, when the function S is obtained in this manner, we have

$$\frac{\partial S}{\partial q_i} = \left(\frac{\partial S}{\partial q_i}\right)_\alpha + \sum_k \left(\frac{\partial S}{\partial \alpha_k}\right)_q \frac{\partial \alpha_k}{\partial q_i} = \left(\frac{\partial S}{\partial q_i}\right)_\alpha.$$

The quantities $(\partial S/\partial q_i)_\alpha$ satisfy the Hamilton–Jacobi equation, since the function $S(t, q; \alpha)$ is assumed to be a complete integral of that equation. The quantities $\partial S/\partial q_i$ therefore satisfy the same equation.

Thus the solution of the problem of the motion of a mechanical system by the Hamilton–Jacobi method proceeds as follows. From the Hamiltonian, we form the Hamilton–Jacobi equation, and find its complete integral (47.2). Differentiating this with respect to the arbitrary constants α and equating the derivatives to new constants β, we obtain s algebraic equations

$$\partial S/\partial \alpha_i = \beta_i, \tag{47.4}$$

whose solution gives the co-ordinates q as functions of time and of the $2s$ arbitrary constants. The momenta as functions of time may then be found from the equations $p_i = \partial S/\partial q_i$.

If we have an incomplete integral of the Hamilton–Jacobi equation, depending on fewer than s arbitrary constants, it cannot give the general integral of the equations of motion, but it can be used to simplify the finding of the general integral. For example, if a function S involving one arbitrary constant α is known, the relation $\partial S/\partial \alpha = $ constant gives one equation between $q_1, ..., q_s$ and t.

The Hamilton–Jacobi equation takes a somewhat simpler form if the function H does not involve the time explicitly, i.e. if the system is conservative. The time-dependence of the action is given by a term $-Et$:

$$S = S_0(q) - Et \tag{47.5}$$

(see §44), and substitution in (47.1) gives for the abbreviated action $S_0(q)$ the Hamilton–Jacobi equation in the form

$$H\left(q_1, ..., q_s; \frac{\partial S_0}{\partial q_1}, ..., \frac{\partial S_0}{\partial q_s}\right) = E. \tag{47.6}$$

§48. Separation of the variables

In a number of important cases, a complete integral of the Hamilton–Jacobi equation can be found by "separating the variables", a name given to the following method.

Let us assume that some co-ordinate, q_1 say, and the corresponding derivative $\partial S/\partial q_1$ appear in the Hamilton–Jacobi equation only in some combination $\phi(q_1, \partial S/\partial q_1)$ which does not involve the other co-ordinates, time, or derivatives, i.e. the equation is of the form

$$\Phi\left\{q_i, t, \frac{\partial S}{\partial q_i}, \frac{\partial S}{\partial t}, \phi\left(q_1, \frac{\partial S}{\partial q_1}\right)\right\} = 0, \tag{48.1}$$

where q_i denotes all the co-ordinates except q_1.

We seek a solution in the form of a sum:

$$S = S'(q_i, t) + S_1(q_1); \tag{48.2}$$

substituting this in equation (48.1), we obtain

$$\Phi\left\{q_i, t, \frac{\partial S'}{\partial q_i}, \frac{\partial S'}{\partial t}, \phi\left(q_1, \frac{dS_1}{dq_1}\right)\right\} = 0. \tag{48.3}$$

Let us suppose that the solution (48.2) has been found. Then, when it is substituted in equation (48.3), the latter must become an identity, valid (in particular) for any value of the co-ordinate q_1. When q_1 changes, only the function ϕ is affected, and so, if equation (48.3) is an identity, ϕ must be a constant. Thus equation (48.3) gives the two equations

$$\phi(q_1, dS_1/dq_1) = \alpha_1, \tag{48.4}$$

$$\Phi\{q_i, t, \partial S'/\partial q_i, \partial S'/\partial t, \alpha_1\} = 0, \tag{48.5}$$

where α_1 is an arbitrary constant. The first of these is an ordinary differential equation, and the function $S_1(q_1)$ is obtained from it by simple integration. The remaining partial differential equation (48.5) involves fewer independent variables.

If we can successively separate in this way all the s co-ordinates and the time, the finding of a complete integral of the Hamilton–Jacobi equation is reduced to quadratures. For a conservative system we have in practice to separate only s variables (the co-ordinates) in equation (47.6), and when this separation is complete the required integral is

$$S = \sum_k S_k(q_k; \alpha_1, \alpha_2, ..., \alpha_s) - E(\alpha_1, ..., \alpha_s)t, \tag{48.6}$$

where each of the functions S_k depends on only one co-ordinate; the energy E, as a function of the arbitrary constants $\alpha_1, ..., \alpha_s$, is obtained by substituting $S_0 = \Sigma S_k$ in equation (47.6).

A particular case is the separation of a cyclic variable. A cyclic co-ordinate q_1 does not appear explicitly in the Hamiltonian, nor therefore in the Hamilton–Jacobi equation. The function $\phi(q_1, \partial S/\partial q_1)$ reduces to $\partial S/\partial q_1$ simply, and from equation (48.4) we have simply $S_1 = \alpha_1 q_1$, so that

$$S = S'(q_i, t) + \alpha_1 q_1. \tag{48.7}$$

The constant α_1 is just the constant value of the momentum $p_1 = \partial S/\partial q_1$ corresponding to the cyclic co-ordinate.

The appearance of the time in the term $-Et$ for a conservative system corresponds to the separation of the "cyclic variable" t.

Thus all the cases previously considered of the simplification of the integration of the equations of motion by the use of cyclic variables are embraced by the method of separating the variables in the Hamilton–Jacobi equation. To those cases are added others in which the variables can be separated even though they are not cyclic. The Hamilton–Jacobi treatment is consequently the most powerful method of finding the general integral of the equations of motion.

To make the variables separable in the Hamilton–Jacobi equation the co-ordinates must be appropriately chosen. We shall consider some examples of separating the variables in different co-ordinates, which may be of physical interest in connection with problems of the motion of a particle in various external fields.

(1) *Spherical co-ordinates.* In these co-ordinates (r, θ, ϕ), the Hamiltonian is

$$H = \frac{1}{2m}\left(p_r^2 + \frac{p_\theta^2}{r^2} + \frac{p_\phi^2}{r^2 \sin^2\theta}\right) + U(r, \theta, \phi),$$

and the variables can be separated if

$$U = a(r) + \frac{b(\theta)}{r^2} + \frac{c(\phi)}{r^2 \sin^2\theta},$$

where $a(r)$, $b(\theta)$, $c(\phi)$ are arbitrary functions. The last term in this expression for U is unlikely to be of physical interest, and we shall therefore take

$$U = a(r) + b(\theta)/r^2. \tag{48.8}$$

In this case the Hamilton–Jacobi equation for the function S_0 is

$$\frac{1}{2m}\left(\frac{\partial S_0}{\partial r}\right)^2 + a(r) + \frac{1}{2mr^2}\left[\left(\frac{\partial S_0}{\partial \theta}\right)^2 + 2mb(\theta)\right] + \frac{1}{2mr^2\sin^2\theta}\left(\frac{\partial S_0}{\partial \phi}\right)^2 = E.$$

Since the co-ordinate ϕ is cyclic, we seek a solution in the form $S_0 = p_\phi\phi + S_1(r) + S_2(\theta)$, obtaining for the functions $S_1(r)$ and $S_2(\theta)$ the equations

$$\left(\frac{dS_2}{d\theta}\right)^2 + 2mb(\theta) + \frac{p_\phi^2}{\sin^2\theta} = \beta,$$

$$\frac{1}{2m}\left(\frac{dS_1}{dr}\right)^2 + a(r) + \frac{\beta}{2mr^2} = E.$$

Integration gives finally

$$S = -Et + p_\phi\phi + \int\sqrt{[\beta - 2mb(\theta) - p_\phi^2/\sin^2\theta]}\,d\theta + \\ + \int\sqrt{\{2m[E - a(r)] - \beta/r^2\}}\,dr. \tag{48.9}$$

The arbitrary constants in (48.9) are p_ϕ, β and E; on differentiating with respect to these and equating the results to other constants, we have the general solution of the equations of motion.

(2) *Parabolic co-ordinates.* The passage from cylindrical co-ordinates (here denoted by ρ, ϕ, z) to parabolic co-ordinates ξ, η, ϕ is effected by the formulae

$$z = \tfrac{1}{2}(\xi - \eta), \qquad \rho = \sqrt{(\xi\eta)}. \tag{48.10}$$

The co-ordinates ξ and η take values from 0 to ∞; the surfaces of constant ξ and η are easily seen to be two families of paraboloids of revolution, with the z-axis as the axis of symmetry. The equations (48.10) can also be written, in terms of

$$r = \sqrt{(z^2+\rho^2)} = \tfrac{1}{2}(\xi+\eta) \tag{48.11}$$

(i.e. the radius in spherical co-ordinates), as

$$\xi = r+z, \qquad \eta = r-z. \tag{48.12}$$

Let us now derive the Lagrangian of a particle in the co-ordinates ξ, η, ϕ. Differentiating the expressions (48.10) with respect to time and substituting in the Lagrangian in cylindrical co-ordinates

$$L = \tfrac{1}{2}m(\rho^2+\rho^2\dot\phi^2+\dot z^2) - U(\rho,\phi,z),$$

we obtain

$$L = \tfrac{1}{8}m(\xi+\eta)\left(\frac{\dot\xi^2}{\xi}+\frac{\dot\eta^2}{\eta}\right)+\tfrac{1}{2}m\xi\eta\dot\phi^2 - U(\xi,\eta,\phi). \tag{48.13}$$

The momenta are $p_\xi = \tfrac{1}{4}m(\xi+\eta)\dot\xi/\xi$, $p_\eta = \tfrac{1}{4}m(\xi+\eta)\dot\eta/\eta$, $p_\phi = m\xi\eta\dot\phi$, and the Hamiltonian is

$$H = \frac{2}{m}\frac{\xi p_\xi^2+\eta p_\eta^2}{\xi+\eta}+\frac{p_\phi^2}{2m\xi\eta}+ U(\xi,\eta,\phi). \tag{48.14}$$

The physically interesting cases of separable variables in these co-ordinates correspond to a potential energy of the form

$$U = \frac{a(\xi)+b(\eta)}{\xi+\eta} = \frac{a(r+z)+b(r-z)}{2r}. \tag{48.15}$$

The equation for S_0 is

$$\frac{2}{m(\xi+\eta)}\left[\xi\left(\frac{\partial S_0}{\partial\xi}\right)^2+\eta\left(\frac{\partial S_0}{\partial\eta}\right)^2\right]+\frac{1}{2m\xi\eta}\left(\frac{\partial S_0}{\partial\phi}\right)^2+\frac{a(\xi)+b(\eta)}{\xi+\eta} = E.$$

The cyclic co-ordinate ϕ can be separated as a term $p_\phi\phi$. Multiplying the equation by $m(\xi+\eta)$ and rearranging, we then have

$$2\xi\left(\frac{\partial S_0}{\partial\xi}\right)^2+ma(\xi)-mE\xi+\frac{p_\phi^2}{2\xi}+2\eta\left(\frac{\partial S_0}{\partial\eta}\right)^2+mb(\eta)-mE\eta+\frac{p_\phi^2}{2\eta} = 0.$$

Putting $S_0 = p_\phi\phi+S_1(\xi)+S_2(\eta)$, we obtain the two equations

$$2\xi\left(\frac{dS_1}{d\xi}\right)^2+ma(\xi)-mE\xi+\frac{p_\phi^2}{2\xi} = \beta,$$

$$2\eta\left(\frac{dS_2}{d\eta}\right)^2+mb(\eta)-mE\eta+\frac{p_\phi^2}{2\eta} = -\beta,$$

integration of which gives finally

$$S = -Et + p_\phi\phi + \int \sqrt{\left[\tfrac{1}{2}mE + \frac{\beta}{2\xi} - \frac{ma(\xi)}{2\xi} - \frac{p_\phi{}^2}{4\xi^2} \right]} \, d\xi +$$

$$+ \int \sqrt{\left[\tfrac{1}{2}mE - \frac{\beta}{2\eta} - \frac{mb(\eta)}{2\eta} - \frac{p_\phi{}^2}{4\eta^2} \right]} \, d\eta. \qquad (48.16)$$

Here the arbitrary constants are p_ϕ, β and E.

(3) *Elliptic co-ordinates.* These are ξ, η, ϕ, defined by

$$\rho = \sigma\sqrt{[(\xi^2-1)(1-\eta^2)]}, \qquad z = \sigma\xi\eta. \qquad (48.17)$$

The constant σ is a parameter of the transformation. The co-ordinate ξ takes values from 1 to ∞, and η from -1 to $+1$. The definitions which are geometrically clearest† are obtained in terms of the distances r_1 and r_2 to points A_1 and A_2 on the z-axis for which $z = \pm\sigma$: $r_1 = \sqrt{[(z-\sigma)^2+\rho^2]}$, $r_2 = \sqrt{[(z+\sigma)^2+\rho^2]}$. Substitution of (48.17) gives

$$r_1 = \sigma(\xi-\eta), \qquad r_2 = \sigma(\xi+\eta),$$
$$\xi = (r_2+r_1)/2\sigma, \qquad \eta = (r_2-r_1)/2\sigma. \qquad (48.18)$$

Transforming the Lagrangian from cylindrical to elliptic co-ordinates, we find

$$L = \tfrac{1}{2}m\sigma^2(\xi^2-\eta^2)\left(\frac{\dot{\xi}^2}{\xi^2-1} + \frac{\dot{\eta}^2}{1-\eta^2} \right) +$$
$$+ \tfrac{1}{2}m\sigma^2(\xi^2-1)(1-\eta^2)\dot{\phi}^2 - U(\xi, \eta, \phi). \qquad (48.19)$$

The Hamiltonian is therefore

$$H = \frac{1}{2m\sigma^2(\xi^2-\eta^2)}\left[(\xi^2-1)p_\xi{}^2 + (1-\eta^2)p_\eta{}^2 + \left(\frac{1}{\xi^2-1} + \frac{1}{1-\eta^2} \right)p_\phi{}^2 \right] +$$
$$+ U(\xi, \eta, \phi). \qquad (48.20)$$

The physically interesting cases of separable variables correspond to a potential energy

$$U = \frac{a(\xi)+b(\eta)}{\xi^2-\eta^2} = \frac{\sigma^2}{r_1 r_2}\left\{ a\left(\frac{r_2+r_1}{2\sigma} \right) + b\left(\frac{r_2-r_1}{2\sigma} \right) \right\}, \qquad (48.21)$$

where $a(\xi)$ and $b(\eta)$ are arbitrary functions. The result of separating the variables in the Hamilton–Jacobi equation is

$$S = -Et + p_\phi\phi + \int \sqrt{\left[2m\sigma^2 E + \frac{\beta - 2m\sigma^2 a(\xi)}{\xi^2-1} - \frac{p_\phi{}^2}{(\xi^2-1)^2} \right]} \, d\xi +$$
$$+ \int \sqrt{\left[2m\sigma^2 E - \frac{\beta + 2m\sigma^2 b(\eta)}{1-\eta^2} - \frac{p_\phi{}^2}{(1-\eta^2)^2} \right]} \, d\eta. \qquad (48.22)$$

---

† The surfaces of constant ξ are the ellipsoids $z^2/\sigma^2\xi^2 + \rho^2/\sigma^2(\xi^2-1) = 1$, of which A_1 and A_2 are the foci; the surfaces of constant η are the hyperboloids $z^2/\sigma^2\eta^2 - \rho^2/\sigma^2(1-\eta^2) = 1$, also with foci A_1 and A_2.

PROBLEMS

PROBLEM 1. Find a complete integral of the Hamilton–Jacobi equation for motion of a particle in a field $U = \alpha/r - Fz$ (a combination of a Coulomb field and a uniform field), and find a conserved function of the co-ordinates and momenta that is specific to this motion.

SOLUTION. The field is of the type (48.15), with $a(\xi) = \alpha - \frac{1}{2}F\xi^2$, $b(\eta) = \alpha + \frac{1}{2}F\eta^2$. The complete integral of the Hamilton–Jacobi equation is given by (48.16) with these functions $a(\xi)$ and $b(\eta)$. To determine the significance of the constant β, we write the equations

$$2\xi p_\xi^2 + ma(\xi) - mE\xi + \tfrac{1}{2}p_\phi^2/\xi = \beta,$$

$$2\eta p_\eta^2 + mb(\eta) - mE\eta + \tfrac{1}{2}p_\phi^2/\eta = -\beta.$$

Subtracting, and expressing the momenta $p_\xi = \partial S/\partial \xi$ and $p_\eta = \partial S/\partial \eta$ in terms of the momenta $p_\rho = \partial S/\partial \rho$ and $p_z = \partial S/\partial z$ in cylindrical co-ordinates, we obtain after a simple calculation

$$\beta = - m\left[\frac{\alpha z}{r} + \frac{p_\rho}{m}(zp_\rho - \rho p_z) + \frac{p_\phi^2}{m\rho^2}z\right] - \tfrac{1}{2}mF\rho^2.$$

The expression in the brackets is an integral of the motion that is specific to the pure Coulomb field (the z-component of the vector (15.17)).

PROBLEM 2. The same as Problem 1, but for a field $U = \alpha_1/r_1 + \alpha_2/r_2$ (the Coulomb field of two fixed points at a distance 2σ apart).

SOLUTION. This field is of the type (48.21), with $a(\xi) = (\alpha_1 + \alpha_2)\xi/\sigma$, $b(\eta) = (\alpha_1 - \alpha_2)\eta/\sigma$. The action $S(\xi, \eta, \varphi, t)$ is obtained by substituting these expressions in (48.22). The significance of the constant β is found in a manner similar to that in Problem 1; in this case it expresses the conservation of the quantity

$$\beta = \sigma^2\left(p^2 + \frac{p_\phi^2}{\rho^2}\right) - M^2 + 2m\sigma(\alpha_1\cos\theta_1 + \alpha_2\cos\theta_2),$$

$$M^2 = (\mathbf{r} \times \mathbf{p})^2$$

$$= p^2z^2 + p_z^2\rho^2 + \frac{r^2p_\phi^2}{\rho^2} - 2z\rho p_z p_\rho,$$

and θ_1 and θ_2 are the angles shown in Fig. 55.

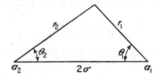

FIG. 55

§49. Adiabatic invariants

Let us consider a mechanical system executing a finite motion in one dimension and characterised by some parameter λ which specifies the properties of the system or of the external field in which it is placed,† and let us suppose that λ varies slowly (*adiabatically*) with time as the result of some external action; by a "slow" variation we mean one in which λ varies only slightly during the period T of the motion:

$$T\,d\lambda/dt \ll \lambda. \tag{49.1}$$

† To simplify the formulae, we assume that there is only one such parameter, but all the results remain valid for any number of parameters.

If λ were constant, the system would be closed and would execute a strictly periodic motion with a constant energy E and a fixed period $T(E)$. When the parameter λ is variable, the system is not closed and its energy is not conserved. However, since λ is assumed to vary only slowly, the rate of change $\dot{E}$ of the energy will also be small. If this rate is averaged over the period T and the "rapid" oscillations of its value are thereby smoothed out, the resulting value $\dot{E}$ determines the rate of steady slow variation of the energy of the system, and this rate will be proportional to the rate of change $\dot{\lambda}$ of the parameter. In other words, the slowly varying quantity E, taken in this sense, will behave as some function of λ. The dependence of E on λ can be expressed as the constancy of some combination of E and λ. This quantity, which remains constant during the motion of a system with slowly varying parameters, is called an *adiabatic invariant*.

Let $H(q, p; \lambda)$ be the Hamiltonian of the system, which depends on the parameter λ. According to formula (40.5), the rate of change of the energy of the system is

$$\frac{dE}{dt} = \frac{\partial H}{\partial t} = \frac{\partial H}{\partial \lambda}\frac{d\lambda}{dt}. \tag{49.2}$$

The expression on the right depends not only on the slowly varying quantity λ but also on the rapidly varying quantities q and p. To ascertain the steady variation of the energy we must, according to the above discussion, average (49.2) over the period of the motion. Since λ and therefore $\dot{\lambda}$ vary only slowly, we can take $\dot{\lambda}$ outside the averaging:

$$\frac{\overline{dE}}{dt} = \frac{d\lambda}{dt}\frac{\overline{\partial H}}{\partial \lambda}, \tag{49.3}$$

and in the function $\partial H/\partial \lambda$ being averaged we can regard only q and p, and not λ, as variable. In other words, the averaging is taken over the motion which would occur if λ remained constant.

The averaging may be explicitly written

$$\frac{\overline{\partial H}}{\partial \lambda} = \frac{1}{T}\int_0^T \frac{\partial H}{\partial \lambda}\, dt.$$

According to Hamilton's equation $\dot{q} = \partial H/\partial p$, or $dt = dq \div (\partial H/\partial p)$. The integration with respect to time can therefore be replaced by one with respect to the co-ordinate, with the period T written as

$$T = \int_0^T dt = \oint dq \div (\partial H/\partial p); \tag{49.4}$$

here the $\oint$ sign denotes an integration over the complete range of variation ("there and back") of the co-ordinate during the period.† Thus (49.3) becomes

$$\frac{dE}{dt} = \frac{d\lambda}{dt} \frac{\oint (\partial H/\partial \lambda)\, dq/(\partial H/\partial p)}{\oint dq/(\partial H/\partial p)}. \tag{49.5}$$

As has already been mentioned, the integrations in this formula must be taken over the path for a given constant value of λ. Along such a path the Hamiltonian has a constant value E, and the momentum is a definite function of the variable co-ordinate q and of the two independent constant parameters E and λ. Putting therefore $p = p(q; E, \lambda)$ and differentiating with respect to λ the equation $H(q,p,\lambda) = E$, we have $\partial H/\partial \lambda + (\partial H/\partial p)(\partial p/\partial \lambda) = 0$, or

$$\frac{\partial H/\partial \lambda}{\partial H/\partial p} = -\frac{\partial p}{\partial \lambda}.$$

Substituting this in the numerator of (49.5) and writing the integrand in the denominator as $\partial p/\partial E$, we obtain

$$\frac{dE}{dt} = -\frac{d\lambda \oint (\partial p/\partial \lambda)\, dq}{dt \oint (\partial p/\partial E)\, dq}$$

or

$$\oint \left(\frac{\partial p}{\partial E}\frac{dE}{dt} + \frac{\partial p}{\partial \lambda}\frac{d\lambda}{dt} \right) dq = 0.$$

Finally, this may be written as

$$dI/dt = 0, \tag{49.6}$$

where

$$I \equiv \oint p\, dq/2\pi, \tag{49.7}$$

the integral being taken over the path for given E and λ. This shows that, in the approximation here considered, I remains constant when the parameter λ varies, i.e. I is an adiabatic invariant.

The quantity I is a function of the energy of the system (and of the parameter λ). The partial derivative with respect to energy determines the period of the motion: from (49.4),

$$2\pi \frac{\partial I}{\partial E} = \oint \frac{\partial p}{\partial E}\, dq = T \tag{49.8}$$

---

† If the motion of the system is a rotation, and the co-ordinate q is an angle of rotation ϕ, the integration with respect to ϕ must be taken over a "complete rotation", i.e. from 0 to 2π.

or

$$\partial E/\partial I = \omega, \tag{49.9}$$

where $\omega = 2\pi/T$ is the vibration frequency of the system.

The integral (49.7) has a geometrical significance in terms of the phase path of the system. In the case considered (one degree of freedom), the phase space reduces to a two-dimensional space (i.e. a plane) with co-ordinates p, q, and the phase path of a system executing a periodic motion is a closed curve in the plane. The integral (49.7) taken round this curve is the area enclosed. It can be written also as the area integral

$$I = \iint dp \, dq/2\pi. \tag{49.10}$$

As an example, let us determine the adiabatic invariant for a one-dimensional oscillator. The Hamiltonian is

$$H = \tfrac{1}{2}p^2/m + \tfrac{1}{2}m\omega^2 q^2, \tag{49.11}$$

where ω is the eigenfrequency of the oscillator. The equation of the phase path is given by the law of conservation of energy $H(p, q) = E$. The path is an ellipse with semi-axes $\sqrt{(2mE)}$ and $\sqrt{(2E/m\omega^2)}$, and its area, divided by 2π, is

$$I = E/\omega. \tag{49.12}$$

The adiabatic invariance of I signifies that, when the parameters of the oscillator vary slowly, the energy is proportional to the frequency.

§50. Canonical variables

Now let the parameter λ be constant, so that the system in question is closed. Let us effect a canonical transformation of the variables q and p, taking I as the new "momentum". The generating function is the abbreviated action S_0, expressed as a function of q and I. For S_0 is defined as the integral

$$S_0(q, E; \lambda) = \int p(q, E; \lambda) \, dq, \tag{50.1}$$

taken for a given energy E and parameter λ. For a closed system, however, I is a function of the energy alone, and so S_0 can equally well be written as a function $S_0(q, I; \lambda)$, and the partial derivative $(\partial S_0/\partial q)_E$ is the same as the derivative $(\partial S_0/\partial q)_I$ for constant I. Hence

$$p = \partial S_0(q, I; \lambda)/\partial q, \tag{50.2}$$

corresponding to the first of the formulae (45.8) for a canonical transformation. The second of these formulae gives the new "co-ordinate", which we denote by w:

$$w = \partial S_0(q, I; \lambda)/\partial I. \tag{50.3}$$

The variables I and w are called *canonical variables*; I is called the *action variable* and w the *angle variable*.

Since the generating function $S_0(q, I; \lambda)$ does not depend explicitly on time, the new Hamiltonian H' is just H expressed in terms of the new variables. In other words, H' is the energy $E(I)$, expressed as a function of the action variable. Accordingly, Hamilton's equations in canonical variables are

$$\dot{I} = 0, \qquad \dot{w} = \mathrm{d}E(I)/\mathrm{d}I. \qquad (50.4)$$

The first of these shows that I is constant, as it should be; the energy is constant, and I is so too. From the second equation we see that the angle variable is a linear function of time:

$$w = \frac{\mathrm{d}E}{\mathrm{d}I}t + \text{constant} = \omega(I)t + \text{constant}; \qquad (50.5)$$

it is the phase of the oscillations.

The action $S_0(q, I)$ is a many-valued function of the co-ordinates. During each period this function increases by

$$\Delta S_0 = 2\pi I, \qquad (50.6)$$

as is evident from (50.1) and the definition of I (49.7). During the same time the angle variable increases by

$$\Delta w = \Delta(\partial S_0/\partial I) = \partial(\Delta S_0)/\partial I = 2\pi. \qquad (50.7)$$

Conversely, if we express q and p, or any one-valued function $F(q, p)$ of them, in terms of the canonical variables, then they remain unchanged when w increases by 2π (with I constant). That is, any one-valued function $F(q, p)$, when expressed in terms of the canonical variables, is a periodic function of w with period 2π.

The equations of motion can also be formulated in canonical variables for a system that is not closed, in which the parameter λ is time-dependent. The transformation to these variables is again effected by formulae (50.2), (50.3), with a generating function S_0 given by the integral (50.1) and expressed in terms of the variable I given by the integral (49.7). The indefinite integral (50.1) and the definite integral (49.7) are calculated as if the parameter $\lambda(t)$ had a given fixed value; that is, $S_0(q, I; \lambda(t))$ is the previous function with the constant λ finally replaced by the specified function $\lambda(t)$.[†]

Since the generating function is now, like the parameter λ, an explicit function of the time, the new Hamiltonian H' is different from the old one, which was the energy $E(I)$. According to the general formulae of the canonical transformation (45.8), we have

$$H' = E(I; \lambda) + \partial S_0/\partial t$$
$$= E(I; \lambda) + \Lambda\dot{\lambda}, \qquad (50.8)$$

[†] It must be emphasised, however, that the function S_0 thus determined is not the true abbreviated action for a system with a time-dependent Hamiltonian.

with the notation

$$\Lambda = (\partial S_0/\partial \lambda)_{q,I};$$ (50.9)

here Λ must be expressed in terms of I and w by (50.3) after the differentiation with respect to λ.

Hamilton's equations now become

$$\dot{I} = -\frac{\partial H'}{\partial w} = -\left(\frac{\partial \Lambda}{\partial w}\right)_{I,\lambda} \dot{\lambda},$$ (50.10)

$$\dot{w} = \frac{\partial H'}{\partial I} = \omega(I;\lambda) + \left(\frac{\partial \Lambda}{\partial I}\right)_{w,\lambda} \dot{\lambda},$$ (50.11)

where $\omega = (\partial E/\partial I)_\lambda$ is the oscillation frequency, again calculated as if λ were constant.

PROBLEM

Write down the equations of motion in canonical variables for a harmonic oscillator (whose Hamiltonian is (49.11)) with time-dependent frequency.

SOLUTION. Since all the operations in (50.1)–(50.3) are for constant λ (λ being in this case the frequency ω itself), the relation of q and p to w has the same form as for constant frequency with $w = \omega t$:

$$q = \sqrt{\frac{2E}{m\omega^2}} \sin w = \sqrt{\frac{2I}{m\omega}} \sin w,$$

$$p = \sqrt{(2I\omega m)} \cos w.$$

Hence

$$S_0 = \int p \, dq = \int p(\partial q/\partial w)_{I,\omega} \, dw = 2I \int \cos^2 w \, dw$$

and

$$\Lambda = \left(\frac{\partial S_0}{\partial \omega}\right)_{q,I} = \left(\frac{\partial S_0}{\partial w}\right)_I \left(\frac{\partial w}{\partial \omega}\right)_q = \frac{I}{2\omega} \sin 2w.$$

Equations (50.10) and (50.11) then become

$$\dot{I} = -I(\dot{\omega}/\omega) \cos 2w, \quad \dot{w} = \omega + (\dot{\omega}/2\omega) \sin 2w.$$

§51. Accuracy of conservation of the adiabatic invariant

The equation of motion in the form (50.10) allows a further proof that the action variable is an adiabatic invariant.

The function $S_0(q, I; \lambda)$ is not a single-valued function of q: when the co-ordinate returns to its original value, S_0 increases by an integral multiple of $2\pi I$. The derivative (50.9), however, *is* single-valued, since the differentiation is at constant I and the increments of S_0 disappear. The function Λ, like any single-valued function, is a periodic function when expressed in terms of the angle variable w. The mean value, over the period, of the derivative $\partial \Lambda/\partial w$ of a periodic function is zero. Hence, on averaging (50.10) and taking λ outside the mean value (when λ varies only slowly), we have

$$\bar{\dot{I}} = -\overline{(\partial \Lambda/\partial w)_I} \, \dot{\lambda} = 0,$$ (51.1)

as was to be proved.

The equations of motion (50.10) and (50.11) enable us to consider the accuracy with which the adiabatic invariant is conserved. The question may be stated as follows: let the parameter $\lambda(t)$ tend to constant limits λ_- and λ_+ as $t \to -\infty$ and $t \to +\infty$; given the initial $(t \to -\infty)$ value I_- of the adiabatic invariant, find the change in it, $\Delta I = I_+ - I_-$ as $t \to +\infty$.

From (50.10),

$$\Delta I = -\int_{-\infty}^{\infty} \frac{\partial \Lambda}{\partial w} \dot{\lambda}\, dt. \tag{51.2}$$

As shown above, Λ is a periodic function of w, with period 2π; let us expand it as a Fourier series

$$\Lambda = \sum_{l=-\infty}^{\infty} e^{ilw} \Lambda_l. \tag{51.3}$$

Since Λ is real, the expansion coefficients are such that $\Lambda_{-l} = \Lambda_l^*$. Hence

$$\frac{\partial \Lambda}{\partial w} = \sum_{l=-\infty}^{\infty} ile^{ilw} \Lambda_l$$

$$= 2\,\mathrm{re}\sum_{l=1}^{\infty} ile^{ilw} \Lambda_l. \tag{51.4}$$

When $\dot{\lambda}$ is sufficiently small, $\dot{w}$ is positive (its sign being the same as that of ω; see (50.11)), i.e. w is a monotonic function of the time t. When we change from integration over t to integration over w in (51.2), the limits are unaltered:

$$\Delta I = -\int_{-\infty}^{\infty} \frac{\partial \Lambda}{\partial w} \frac{d\lambda}{dt} \frac{dt}{dw}\, dw. \tag{51.5}$$

Substituting (51.4), we can transform the integral by formally treating w as a complex variable. We assume that the integrand has no singularities for real w, and displace the path of integration off the real axis into the upper half-plane of this complex variable. The contour is then "caught up" at the singularities of the integrand, and forms loops round them, as shown schematically in Fig. 56. Let w_0 be the singularity nearest the real axis, i.e.

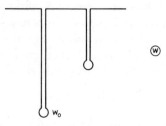

Fig. 56

the one with the smallest (positive) imaginary part. The principal contribution to the integral (51.5) comes from the neighbourhood of this point, and each term in the series (51.4) gives a contribution containing a factor $\exp(-l\operatorname{im} w_0)$. Again retaining only the term with the negative exponent of smallest magnitude (i.e. the term with $l = 1$), we find†

$$\Delta I \sim \exp(-\operatorname{im} w_0). \tag{51.6}$$

Let t_0 be the (complex) "instant" corresponding to the singularity w_0: $w(t_0) = w_0$. In general, $|t_0|$ has the same order of magnitude as the characteristic time τ of variation of the parameters of the system.‡ The order of magnitude of the exponent in (51.6) is

$$\operatorname{im} w_0 \sim \omega\tau \sim \tau/T. \tag{51.7}$$

Since we assume that $\tau \gg T$, this exponent is large. Thus the difference $I_+ - I_-$ decreases exponentially as the rate of variation of the parameters of the system decreases.||

To determine w_0 in the first approximation with respect to T/τ (i.e. retaining only the term $\sim(T/\tau)^{-1}$ in the exponent), we can omit from (50.11) the small term in $\dot{\lambda}$:

$$dw/dt = \omega(I, \lambda(t)), \tag{51.8}$$

and the argument I of the function $\omega(I, \lambda)$ is taken to have a constant value, say I_-. Then

$$w_0 = \int^{t_0} \omega(I, \lambda(t))\, dt; \tag{51.9}$$

the lower limit may be taken as any real value of t, since it does not affect the required imaginary part of w_0.

The integral (51.5) with w from (51.8) (and with one term from the series (51.4) as $\partial\Lambda/\partial w$) becomes

$$\Delta I \sim \operatorname{re} \int i e^{iw} \frac{\lambda\, dw}{\omega(I, \lambda)}. \tag{51.10}$$

Hence we see that the singularities that are in question as regards the nearest to the real axis are the singularities (poles and branch points) of the functions

---

† In special cases it may happen that the expansion (51.4) does not include a term with $l = 1$ (see, for example, the Problem at the end of this section); in every case, we must take the term with the lowest value of l present in the series.

‡ If the slowness of variation of the parameter λ is expressed by its depending on t only through a ratio $\xi = t/\tau$ with τ large, then $t_0 = \tau\xi_0$, where ξ_0 is a singularity of $\lambda(\xi)$ that is independent of τ.

|| Note that, if the initial and final values of $\lambda(t)$ are the same ($\lambda_+ = \lambda_-$), then not only the difference ΔI but also the difference $\Delta E = E_+ - E_-$ of the final and initial energies are exponentially small; from (49.9), $\Delta E = \omega \Delta I$ in that case.

$\dot{\lambda}(t)$ and $1/\omega(t)$. Here it should be remembered that the conclusion that ΔI is exponentially small depends on the hypothesis that these functions have no real singularities.

PROBLEM

Estimate ΔI for a harmonic oscillator with a frequency that varies slowly according to

$$\omega^2 = \omega_0{}^2 \frac{1 + ae^{\alpha t}}{1 + e^{\alpha t}}$$

from $\omega_- = \omega_0$ for $t = -\infty$ to $\omega_+ = \sqrt{a}\,\omega_0$ for $t = \infty$ $(a > 0,\ \alpha \ll \omega_0)$.†

SOLUTION. Taking as the parameter λ the frequency ω itself, we have

$$\frac{\dot{\lambda}}{\omega} = \tfrac{1}{2}\alpha\left(\frac{a}{e^{-\alpha t} + a} - \frac{1}{e^{-\alpha t} + 1}\right).$$

This function has poles for $e^{-\alpha t} = -1$ and $e^{-\alpha t} = -a$. Calculating the integral $\int \omega\, dt$, we find that the smallest value of im w_0 comes from one of the poles $\alpha t_0 = -\log(-a)$, and is

$$\operatorname{im} w_0 = \omega_0 \pi/\alpha \text{ for } a > 1,$$
$$= \omega_0 \pi \sqrt{a}/\alpha \text{ for } a < 1.$$

For a harmonic oscillator, $\Lambda \sim \sin 2w$ (see §50, Problem), so that the series (51.3) reduces to two terms (with $l = \pm 2$). Thus, for a harmonic oscillator,

$$\Delta I \sim \exp(-2 \operatorname{im} w_0).$$

§52. Conditionally periodic motion

Let us consider a system with any number of degrees of freedom, executing a motion finite in all the co-ordinates, and assume that the variables can be completely separated in the Hamilton–Jacobi treatment. This means that, when the co-ordinates are appropriately chosen, the abbreviated action can be written in the form

$$S_0 = \sum_i S_i(q_i), \tag{52.1}$$

as a sum of functions each depending on only one co-ordinate.

Since the generalised momenta are $p_i = \partial S_0/\partial q_i = dS_i/dq_i$, each function S_i can be written

$$S_i = \int p_i\, dq_i. \tag{52.2}$$

These are many-valued functions. Since the motion is finite, each co-ordinate can take values only in a finite range. When q_i varies "there and back" in this range, the action increases by

$$\Delta S_0 = \Delta S_i = 2\pi I_i, \tag{52.3}$$

---

† The harmonic nature of the oscillator is shown by the fact that the oscillation frequency is independent of the energy.

where

$$I_i \equiv \oint p_i \, dq_i/2\pi, \tag{52.4}$$

the integral being taken over the variation of q_i just mentioned.†

Let us now effect a canonical transformation similar to that used in §50, for the case of a single degree of freedom. The new variables are "action variables" I_i and "angle variables"

$$w_i = \partial S_0(q, I)/\partial I_i = \sum_k \partial S_k(q_k, I)/\partial I_i, \tag{52.5}$$

where the generating function is again the action expressed as a function of the co-ordinates and the I_i. The equations of motion in these variables are $\dot{I}_i = 0$, $\dot{w}_i = \partial E(I)/\partial I_i$, which give

$$I_i = \text{constant}, \tag{52.6}$$

$$w_i = [\partial E(I)/\partial I_i]t + \text{constant}. \tag{52.7}$$

We also find, analogously to (50.7), that a variation "there and back" of the co-ordinate q_i corresponds to a change of 2π in w_i:

$$\Delta w_i = 2\pi. \tag{52.8}$$

In other words, the quantities $w_i(q, I)$ are many-valued functions of the co-ordinates: when the latter vary and return to their original values, the w_i may vary by any integral multiple of 2π. This property may also be formulated as a property of the function $w_i(p, q)$, expressed in terms of the co-ordinates and momenta, in the phase space of the system. Since the I_i, expressed in terms of p and q, are one-valued functions, substitution of $I_i(p, q)$ in $w_i(q, I)$ gives a function $w_i(p, q)$ which may vary by any integral multiple of 2π (including zero) on passing round any closed path in phase space.

Hence it follows that any one-valued function‡ $F(p, q)$ of the state of the system, if expressed in terms of the canonical variables, is a periodic function of the angle variables, and its period in each variable is 2π. It can be expanded as a multiple Fourier series:

$$F = \sum_{l_1=-\infty}^{\infty} \cdots \sum_{l_s=-\infty}^{\infty} A_{l_1 l_2 \cdots l_s} \exp\{i(l_1 w_1 + \dots + l_s w_s)\},$$

---

† It should be emphasised, however, that this refers to the formal variation of the co-ordinate q_i over the whole possible range of values, not to its variation during the period of the actual motion as in the case of motion in one dimension. An actual finite motion of a system with several degrees of freedom not only is not in general periodic as a whole, but does not even involve a periodic time variation of each co-ordinate separately (see below).

‡ Rotational co-ordinates ϕ (see the second footnote to §49) are not in one-to-one relation with the state of the system, since the position of the latter is the same for all values of ϕ differing by an integral multiple of 2π. If the co-ordinates q include such angles, therefore, these can appear in the function $F(p, q)$ only in such expressions as $\cos \phi$ and $\sin \phi$, which are in one-to-one relation with the state of the system.

where $l_1, l_2, ..., l_s$ are integers. Substituting the angle variables as functions of time, we find that the time dependence of F is given by a sum of the form

$$F = \sum_{l_1=-\infty}^{\infty} \cdots \sum_{l_s=-\infty}^{\infty} A_{l_1 l_2 \cdots l_s} \exp\left\{ it\left(l_1 \frac{\partial E}{\partial I_1} + \cdots + l_s \frac{\partial E}{\partial I_s} \right) \right\}. \qquad (52.9)$$

Each term in this sum is a periodic function of time, with frequency

$$l_1 \omega_1 + \cdots + l_s \omega_s, \qquad (52.10)$$

which is a sum of integral multiples of the fundamental frequencies

$$\omega_i = \partial E/\partial I_i. \qquad (52.11)$$

Since the frequencies (52.10) are not in general commensurable, the sum itself is not a periodic function, nor, in particular, are the co-ordinates q and momenta p of the system.

Thus the motion of the system is in general not strictly periodic either as a whole or in any co-ordinate. This means that, having passed through a given state, the system does not return to that state in a finite time. We can say, however, that in the course of a sufficient time the system passes arbitrarily close to the given state. For this reason such a motion is said to be *conditionally periodic*.

In certain particular cases, two or more of the fundamental frequencies ω_i are commensurable for arbitrary values of the I_i. This is called *degeneracy*, and if all s frequencies are commensurable, the motion of the system is said to be *completely degenerate*. In the latter case the motion is evidently periodic, and the path of every particle is closed.

The existence of degeneracy leads, first of all, to a reduction in the number of independent quantities I_i on which the energy of the system depends. If two frequencies ω_1 and ω_2 are such that

$$n_1 \partial E/\partial I_1 = n_2 \partial E/\partial I_2, \qquad (52.12)$$

where n_1 and n_2 are integers, then it follows that I_1 and I_2 appear in the energy only as the sum $n_2 I_1 + n_1 I_2$.

A very important property of degenerate motion is the increase in the number of one-valued integrals of the motion over their number for a general non-degenerate system with the same number of degrees of freedom. In the latter case, of the $2s-1$ integrals of the motion, only s functions of the state of the system are one-valued; these may be, for example, the s quantities I_i. The remaining $s-1$ integrals may be written as differences

$$w_i \partial E/\partial I_k - w_k \partial E/\partial I_i. \qquad (52.13)$$

The constancy of these quantities follows immediately from formula (52.7), but they are not one-valued functions of the state of the system, because the angle variables are not one-valued.

When there is degeneracy, the situation is different. For example, the relation (52.12) shows that, although the integral

$$w_1 n_2 - w_2 n_1 \qquad (52.14)$$

is not one-valued, it is so except for the addition of an arbitrary integral multiple of 2π. Hence we need only take a trigonometrical function of this quantity to obtain a further one-valued integral of the motion.

An example of degeneracy is motion in a field $U = -\alpha/r$ (see Problem). There is consequently a further one-valued integral of the motion (15.17) peculiar to this field, besides the two (since the motion is two-dimensional) ordinary one-valued integrals, the angular momentum M and the energy E, which hold for motion in any central field.

It may also be noted that the existence of further one-valued integrals leads in turn to another property of degenerate motions: they allow a complete separation of the variables for several (and not only one†) choices of the coordinates. For the quantities I_i are one-valued integrals of the motion in co-ordinates which allow separation of the variables. When degeneracy occurs, the number of one-valued integrals exceeds s, and so the choice of those which are the desired I_i is no longer unique.

As an example, we may again mention Keplerian motion, which allows separation of the variables in both spherical and parabolic co-ordinates.

In §49 it has been shown that, for finite motion in one dimension, the action variable is an adiabatic invariant. This statement holds also for systems with more than one degree of freedom. It can be proved, in the general case, by a direct generalisation of the method given at the beginning of §51.

For a multi-dimensional system with a variable parameter $\lambda(t)$, the equations of motion in canonical variables give for the rate of variation of each action variable I_i an expression analogous to (50.10):

$$\dot{I}_i = -\frac{\partial \Lambda}{\partial w_i} \dot{\lambda}_i, \qquad (52.15)$$

where, as before, $\Lambda = (\partial S_0/\partial \lambda)_I$. This equation is to be averaged over a time interval large compared with the fundamental periods of the system but small compared with the time of variation of $\lambda(t)$. The quantity $\dot{\lambda}$ is again taken outside the mean value, and the derivatives $\partial \Lambda/\partial w_i$ are averaged as if the motion took place at constant λ, as a conditionally periodic motion. Then Λ is a unique periodic function of the angle variables w_i, and the mean values of its derivatives $\partial \Lambda/\partial w_i$ are zero.

Finally, we may briefly discuss the properties of finite motion of closed systems with s degrees of freedom in the most general case, where the variables in the Hamilton–Jacobi equation are not assumed to be separable.

---

† We ignore such trivial changes in the co-ordinates as $q_1' = q_1'(q_1)$, $q_2' = q_2'(q_2)$.

The fundamental property of systems with separable variables is that the integrals of the motion I_i, whose number is equal to the number of degrees of freedom, are one-valued. In the general case where the variables are not separable, however, the one-valued integrals of the motion include only those whose constancy is derived from the homogeneity and isotropy of space and time, namely energy, momentum and angular momentum.

The phase path of the system traverses those regions of phase space which are defined by the given constant values of the one-valued integrals of the motion. For a system with separable variables and s one-valued integrals, these conditions define an s-dimensional manifold (hypersurface) in phase space. During a sufficient time, the path of the system passes arbitrarily close to every point on this hypersurface.

In a system where the variables are not separable, however, the number of one-valued integrals is less than s, and the phase path occupies, completely or partly, a manifold of more than s dimensions in phase space.

In degenerate systems, on the other hand, which have more than s integrals of the motion, the phase path occupies a manifold of fewer than s dimensions.

If the Hamiltonian of the system differs only by small terms from one which allows separation of the variables, then the properties of the motion are close to those of a conditionally periodic motion, and the difference between the two is of a much higher order of smallness than that of the additional terms in the Hamiltonian.

PROBLEM

Calculate the action variables for elliptic motion in a field $U = -\alpha/r$.

SOLUTION. In polar co-ordinates r, ϕ in the plane of the motion we have

$$I_\phi = \frac{1}{2\pi} \int_0^{2\pi} p_\phi \, d\phi = M,$$

$$I_r = 2\frac{1}{2\pi} \int_{r_{\min}}^{r_{\max}} \sqrt{\left[2m\left(E + \frac{\alpha}{r}\right) - \frac{M^2}{r^2} \right]} \, dr$$

$$= -M + \alpha\sqrt{(m/2|E|)}.$$

Hence the energy, expressed in terms of the action variables, is $E = -m\alpha^2/2(I_r + I_\phi)^2$. It depends only on the sum $I_r + I_\phi$, and the motion is therefore degenerate; the two fundamental frequencies (in r and in ϕ) coincide.

The parameters p and e of the orbit (see (15.4)) are related to I_r and I_ϕ by

$$p = \frac{I_\phi^2}{m\alpha}, \qquad e^2 = 1 - \left(\frac{I_\phi}{I_\phi + I_r}\right)^2.$$

Since I_r and I_ϕ are adiabatic invariants, when the coefficient α or the mass m varies slowly the eccentricity of the orbit remains unchanged, while its dimensions vary in inverse proportion to α and to m.

INDEX

Acceleration, 1
Action, 2, 138ff.
  abbreviated, 141
  variable, 157
Additivity of
  angular momentum, 19
  energy, 14
  integrals of the motion, 13
  Lagrangians, 4
  mass, 17
  momentum, 15
Adiabatic invariants, 155, 159ff., 165
Amplitude, 59
  complex, 59
Angle variable, 157
Angular momentum, 19ff.
  of rigid body, 105ff.
Angular velocity, 97f.
Area integral, 31n.

Beats, 63
Brackets, Poisson, 135ff.

Canonical equations (VII), 131ff.
Canonical transformation, 143ff.
Canonical variables, 157
Canonically conjugate quantities, 145
Central field, 21, 30
  motion in, 30ff.
Centrally symmetric field, 21
Centre of field, 21
Centre of mass, 17
  system, 41
Centrifugal force, 128
Centrifugal potential, 32, 128
Characteristic equation, 67
Characteristic frequencies, 67
Closed system, 8
Collisions between particles (IV), 41ff.
  elastic, 44ff.
Combination frequencies, 85
Complete integral, 148
Conditionally periodic motion, 164
Conservation laws (II), 13ff.
Conservative systems, 14
Conserved quantities, 13
Constraints, 10
  equations of, 123
  holonomic, 123
Co-ordinates, 1
  cyclic, 30
  generalised, 1ff.
  normal, 68f.

Coriolis force, 128
Couple, 109
Cross-section, effective, for scattering, 49ff.
C system, 41
Cyclic co-ordinates, 30

d'Alembert's principle, 124
Damped oscillations, 74ff.
Damping
  aperiodic, 76
  coefficient, 75
  decrement, 75
Degeneracy, 39, 69, 164f.
  complete, 164
Degrees of freedom, 1
Disintegration of particles, 41ff.
Dispersion-type absorption, 79
Dissipative function, 76f.
Dummy suffix, 99n.

Eccentricity, 36
Eigenfrequencies, 67
Elastic collision, 44
Elliptic functions, 118f.
Elliptic integrals, 26, 118
Energy, 14, 25f.
  centrifugal, 32, 128
  internal, 17
  kinetic, *see* Kinetic energy
  potential, *see* Potential energy
Equations of motion (I), 1ff.
  canonical (VII), 131ff.
  integration of (III), 25ff.
  of rigid body, 107ff.
Eulerian angles, 110ff.
Euler's equations, 115, 119

Finite motion, 25
Force, 9
  generalised, 16
Foucault's pendulum, 129f.
Frame of reference, 4
  inertial, 5f.
  non-inertial, 126ff.
Freedom, degrees of, 1
Frequency, 59
  circular, 59
  combination, 85
Friction, 75, 122

Galilean transformation, 6
Galileo's relativity principle, 6